D1236833

Introduction to Semigroups

Merrill Research and Lecture Series

Erwin Kleinfeld, *Editor*

Introduction to Semigroups

Mario Petrich
Pennsylvania State University

Charles E. Merrill Publishing Company
A Bell & Howell Company
Columbus, Ohio

Published by
Charles E. Merrill Publishing Co.
A Bell & Howell Company
Columbus, Ohio 43216

International Standard Book Number: 0–675–09062–8

Library of Congress Catalog Number: 72–78086

1 2 3 4 5 6 7 8 — 73 74 75 76 77 78 79

Printed in the United States of America

Preface

The theory of semigroups has experienced a vigorous development during the last decades to become an independent branch of algebra with applications to several branches of mathematics.

This monograph grew out of the conviction that this theory has advanced far enough to deserve a book primarily designed for teaching the subject. The problem of choosing the material from the vast information available on various aspects of the theory quickly transformed the original idea of the scope of the monograph to combine it with a reference book for the specialist in semigroup theory as well as for persons interested in branches of algebra related to this theory. This is enhanced by the special point of view adopted, viz., a strong emphasis on structure theorems expressed in terms of the greatest semilattice decomposition and ideal extensions, which appears here for the first time in book form. It is hoped that this monograph will prove a suitable text for the graduate student interested in the subject, a reference book for the specialist, and a useful guide for persons whose interests include this theory.

Most of the material presented here has been first elaborated in a course given to advanced graduate students in 1967–9. I am indebted to students of that class for numerous corrections of and helpful remarks on the original version of this text. Professor B. M. Schein has read the entire material with great care and has freely given of his advice. Professors A. H. Clifford, C. F. Fennemore, T. Pirnot and T. Tamura have suggested several improvements and corrected many errors. The aid of these persons has definitely improved the quality of presentation of the material covered. To these persons and all others who have contributed to the existence of this book, I express my sincere thanks.

Contents

I

Preliminaries

The purpose of this chapter is to introduce basic concepts used throughout this book and to establish their simplest properties. Only the minimum of the notions and properties is discussed here; this is a short preparation of general nature which is needed for more specialized subjects treated in the remaining chapters.

I.1. Introduction

The theory of semigroups had essentially two origins. One was an attempt to generalize both group theory and, to a lesser extent, ring theory to the algebraic system consisting of a single associative operation, which from the group theoretical point of view omits the axioms of the existence of identities and inverses, and from the ring theoretical point of view omits the additive structure of the ring. The second consisted of an algebraic abstraction of the properties of the composition of transformations on a set; this can be understood as a generalization of the abstract treatment of transformation groups. Later sources include an abstraction of certain ideas arising in differential geometry, the theory of automata, transformations on a topological or linear space, etc. From these different origins, a collection of scattered results has emerged which in time have sorted themselves into several, rather separated, branches of what is today called the algebraic theory of semigroups.

For the consolidation and continuity of the separate branches and the theory in general, the first monographs on the subject have exercised a decisive influence: Suškevič's *The theory of generalized groups*, Ljapin's *Semigroups*, and Clifford and Preston's *The algebraic theory of semigroups*. While Suškevič's book is of a more pioneering character and contains mainly the author's own investigations, the latter two monographs try to cover the main aspects of the theory. In Ljapin's book a recurring theme is the appearance of various semigroups of (partial) transformations on a set. In the book by Clifford and Preston a frequently used tool for the study of abstract, and sometimes concrete, semigroups are Green's relations. All of these monographs seem to be primarily written as reference books rather than texts for use in a classroom. Rédei's research tract "The theory of finitely generated commutative semigroups" covers an aspect of this specialized field.

With this book we offer two novelties. The first one is the approach primarily designed as a lecture series to be used in a classroom for the student who should have had some course in modern algebra, say on the senior or first year graduate level in the U.S. Nevertheless, the book is so organized that it can be easily and, we hope successfully, used as a reference. In it the specialist will find many results and topics not covered in the books mentioned above and a few results appearing here for the first time. The second novelty is the mathematical approach to the subject and the choice of the topics covered. Here we adopt the point of view, mostly propagated by Professor T. Tamura and the author, that a semigroup should be studied through its greatest semilattice decomposition. Intuitively the idea consists of decomposing the given semigroup into "smaller subsemigroups," possibly of considerably simpler structure, studying these in detail, and finally studying their mutual relationships within the entire semigroup.

In Chapter I, after this introduction, we define the basic concepts, establish a few properties of these to be used in the succeeding chapters, and discuss some often occurring and most elementary examples. Chapter II deals with the basic tool of our approach, viz., the greatest semilattice decomposition. In Chapter III we consider ideal extensions as these are closely related to the subject started in Chapter II. The theory developed in Chapters II and III is applied in Chapter IV to the class of completely regular semigroups. Finally, in Chapter V, we study the translational hull of an arbitrary and some special semigroups, as the need for this arises from the considerations in Chapters III and IV. This is followed by an appendix which lists certain semigroups of order ≤ 4 and indicates how to construct the remaining ones.

Rather than try to cover a great number of subjects, we have confined ourselves to a small number of topics, which are then mostly covered to the "research boundary." On the one hand, this gives the student a somewhat specialized look at the theory, but, on the other hand, it takes him in a single course from nothing (in the semigroup sense) to a general working knowledge of the subject. This in turn makes it possible for him to do research in any of the areas covered here and makes the reading of other books on semigroups much easier.

It should be mentioned here that besides "abstract" semigroups, i.e., semigroups with only a binary associative operation, treated here, there is a variety of other kinds of semigroups. These are either provided with some additional structure(s), e.g., ordered semigroups, topological semigroups, etc., or are semigroups of various kinds of transformations on algebraic, topological or other structures, e.g., semigroups of operators in functional analysis, or are subsemigroups of a semigroup of particular interest, e.g., semigroups occurring in number theory or the mathematical theory of languages. None of these semigroups will be studied in the present work.

The statements are numbered with 3 numbers, e.g., IV.5.6 means: Chapter IV, Section 5, Statement 6; they are referred to in full except if they appear in the same chapter, in which case the number denoting the chapter is omitted. Each section in Chapters II-V ends with a set of exercises (more difficult ones are starred and some are provided with a hint) and a few references for further reading or for a consultation of the original paper.

I.2. Definition of a Semigroup

The reader should recall the definitions of an ordered pair and the Cartesian product.

I.2.1 DEFINITION. A (*binary*) *operation*, or a *multiplication*, on a set S is a function which to each ordered pair of elements of S associates a unique element of S.

Hence a multiplication on a set S is any function mapping the Cartesian product $S \times S$ into S. Its value at the pair (a,b) will be usually denoted by juxtaposition ab, or sometimes, particularly when several operations on S are considered, by some special symbol such as $a * b$. The notation $(a,b,\ldots \in S)$ stands for "for all $a,b,\ldots$ in S."

I.2.2 DEFINITION. A multiplication on a set S is *associative* if it satisfies the condition

$$(a * b) * c = a * (b * c) \qquad (a,b,c \in S)$$

called the *associative law*. In such a case, the pair $(S,*)$ is a *semigroup*.

It is customary to say "semigroup S" rather than "semigroup $(S,*)$." Here we interchange the meaning of the semigroup and the set on which it is defined even if the multiplication is denoted by a special symbol. This is an established practice and will cause no confusion. The next proposition points to the real significance of the associative law.

I.2.3 PROPOSITION. Every semigroup S satisfies the *general associative law* which says that the value of the product of n elements of S does not depend on the positioning of the parentheses.

Proof. We define first $a_1a_2 \ldots a_n = a_1(a_2(\ldots (a_{n-1}a_n) \ldots))$, and by an inductive argument on the number n of factors show that with any other positioning of parentheses we again obtain the element $a_1a_2 \ldots a_n$. This is trivial for $n = 1,2$ and is given for $n = 3$. Suppose that a is a product of $a_1,a_2, \ldots, a_n$ with $n > 3$ and that the statement holds for all $r < n$. No matter what the positioning of parentheses in a is, we must be able to write $a = bc$ where b is the product of $a_1,a_2, \ldots, a_r$ and c is the product of $a_{r+1}, a_{r+2}, \ldots, a_n$ for some $1 \leq r < n$. By the induction hypothesis, we obtain $b = a_1a_2 \ldots a_r$ and $c = a_{r+1}a_{r+2} \ldots a_n$. Using the induction hypothesis we further have

$$\begin{aligned} a = \mathrm{bc} &= (a_1a_2 \ldots a_r)(a_{r+1}a_{r+2} \ldots a_n) = \\ [a_1(a_2 \ldots a_r)]\,(a_{r+1}a_{r+2} \ldots a_n) &= a_1[(a_2 \ldots a_r)(a_{r+1} \ldots a_n)] \\ &= a_1\,(a_2 \ldots a_ra_{r+1} \ldots a_n) = a_1a_2 \ldots a_n \quad \text{if} \quad r > 1, \\ a = bc &= a_1(a_2 \ldots a_n) = a_1a_2 \ldots a_n \quad \text{if} \quad r = 1. \end{aligned}$$

In light of this proposition, we are able to omit all parentheses from products of an arbitrary number of elements of any semigroups. Nevertheless, we will often use parentheses as an indication of the way a certain expression was derived from the preceding one, which will save explanations.

I.3. Special Subsets of a Semigroup

A number of subsets of a semigroup enjoy special properties relative to the multiplication and are of particular importance.

I.3.1 DEFINITION. A nonempty subset T of a semigroup S is a *subsemigroup* of S if it is closed under the operation of S; i.e., if $a,b \in T$ then $ab \in T$.

It is clear that a subsemigroup of a semigroup with the induced multiplication is a semigroup in its own right.

I.3.2 DEFINITION. A semigroup S is *generated* by its subset G if every element of S can be written as a product of some elements of G. A semigroup S generated by a subset consisting of a single element is a *cyclic* semigroup. If A is a nonempty subset of a semigroup S, then the set

$$\{a_1a_2 \ldots a_n \mid a_i \in A \text{ and } n \text{ is arbitrary}\}$$

is the *subsemigroup of* S *generated by* A.

I.3.3 LEMMA. Let A be a nonempty subset of a semigroup S. Then the subsemigroup of S generated by A is the intersection of all subsemigroups of S containing A.

Proof. Exercise.

Hence the subsemigroup generated by A is under inclusion the least subsemigroup of S containing A.

I.3.4 DEFINITION. A nonempty subset T of a semigroup S is a *left ideal* of S if $s \in S$, $t \in T$ imply $st \in T$; T is a *right ideal* if $s \in S$, $t \in T$ imply $ts \in T$; T is a *two-sided ideal* (or simply an *ideal*) if it is both a left and a right ideal. An ideal of S different from S is a *proper ideal.*

We will often encounter statements pertaining to one-sided concepts, e.g. left ideals. Their analogues pertaining to the other side will not be stated explicitly and will be referred to as "the dual" of the original statement. The following notation will be convenient.

I.3.4 NOTATION. If $A_1, A_2, \ldots, A_n$ are nonempty subsets of a semigroup S, then

$$A_1A_2 \ldots A_n = \{a_1a_2 \ldots a_n \mid a_i \in A_i, \quad 1 \leq i \leq n\}.$$

If $A_i = \{a\}$ is a singleton, then we write $A_1A_2 \ldots A_{i-1}aA_{i+1} \ldots A_n$ and if $A = A_1 = A_2 = \ldots = A_n$, we write A^n instead of $A_1A_2 \ldots A_n$.

In general, we will not make a distinction between a one element set and the single element it contains. The above definitions can be rephrased thus: a nonempty subset T of a semigroup S is a (i) subsemigroup if $T^2 \subseteq T$, (ii) left ideal if $ST \subseteq T$, (iii) right ideal if $TS \subseteq T$, (iv) ideal if $ST \cup TS \subseteq T$. Furthermore, for any nonempty subset A of S, the subsemigroup of S generated by A coincides with $\bigcup_{n=1}^{\infty} A^n$.

I.3.5 LEMMA. Each of the sets (all left ideals, all right ideals, all ideals of a semigroup S) is closed under the following operations: (i) intersection if nonempty, (ii) arbitrary union. In addition, the intersection of a finite number of ideals is an ideal.

Proof. Exercise.

We have seen above that the subsemigroup generated by a set can be equivalently defined as the intersection of all subsemigroups containing that set. We now adopt this approach for ideals.

I.3.6 DEFINITION. The intersection of all left ideals of a semigroup S containing a nonempty subset A of S is the *left ideal generated by A*. A left ideal generated by a one-element set $\{a\}$ is the *principal left ideal generated by a*, and will be denoted by $L(a)$. The corresponding definitions are valid for right ideals with notation $R(a)$, and two-sided ideals with notation $J(a)$.

I.3.7 LEMMA. For any element a of a semigroup S, we have

$$L(a) = a \cup Sa, \qquad R(a) = a \cup aS, \qquad J(a) = a \cup aS \cup Sa \cup SaS;$$

Proof. Exercise.

I.3.8 DEFINITION. A semigroup S is *left* (*right*) *simple* if S is its only left (right) ideal; S is *simple* if S is its only two-sided ideal.

I.3.9 LEMMA. Let S be a semigroup. Then S is left simple if and only if $Sa = S$ for all $a \in S$; S is simple if and only if $SaS = S$ for all $a \in S$.

Proof. Exercise.

Note that this lemma says that S is left simple if and only if for any $a,b \in S$ the equation $xa = b$ has a solution in S, and that S is simple if and only if for any $a,b \in S$ the equation $xay = b$ has a solution in S. It is useful to compare this with the definition of a group involving solvability of such equations.

I.3.10 DEFINITION. The intersection of all ideals of a semigroup S, if nonempty, is the *kernel* of S.

In view of the last statement in 3.5, every finite semigroup has a kernel.

I.3.11 LEMMA. If I is a simple ideal of a semigroup S, then I is the kernel of S.

Proof. If J is an ideal of S, then $I \cap J \supseteq IJ$ and hence $I \cap J$ is nonempty and is thus an ideal of S. But then $I \cap J$ is also an ideal of I, which by simplicity of I implies that $I \cap J = I$ and thus $I \subseteq J$.

I.4. Special Elements of a Semigroup

A number of elements of a semigroup also have special properties relative to the multiplication and play an important role in the study of the subject.

I.4.1 DEFINITION. Let S be a semigroup and s an element of S. An element e of S is a *left identity* of s if $es = s$; a *right identity* of s if $se = s$; a *two-sided identity* of s if $s = es = se$. Furthermore e is a *left* (*right*) *identity* of S if it is a left (right) identity of every element of S; a *two-sided identity* (or simply an *identity*) if it is both a left and a right identity of S.

I.4.2 LEMMA. A semigroup can have at most one identity.

Proof. Exercise.

To any semigroup S, with or without an identity, we can adjoin an identity by taking the set $S \cup e$, where e is some element not contained in S, with multiplication: $s * e = e * s = s$ for all $s \in S \cup e$, and leaving the multiplication among the elements of S unchanged. The following notation will turn out to be quite convenient.

I.4.3 NOTATION. Let S be a semigroup. If S has an identity, set $S^1 = S$, and if S does not have an identity, let S^1 be the semigroup S with an identity adjoined, usually denoted by the symbol 1.

For example, in this notation according to 3.7, $L(a) = S^1a$, $R(a) = aS^1$, $J(a) = S^1aS^1$.

A property of elements of a semigroup, in a certain sense opposite to that of being a left or a right identity, is provided by the following concept.

I.4.4 DEFINITION. An element z of a semigroup S is a *left zero* of S if $zs = z$ for all $s \in S$; a *right zero* if $sz = z$ for all $s \in S$; a *two-sided zero* (or simply a *zero*) if it is both a left and a right zero. If z is a zero of S and $ab = z$ where $a,b \in S$ are both nonzero, then both a and b are *zero divisors*.

I.4.5 LEMMA. A semigroup can have at most one zero.

Proof. Exercise.

Note that z is a left zero of a semigroup S if and only if $\{z\}$ is a right ideal of S. One may define "a left (right, two-sided) zero" of an element analogously as above for identities; however this notion seems to occur somewhat less frequently. A zero can be adjoined to any semigroup S in a fashion analogous to that of adjoining an identity. Indeed, we let 0 be any element not in S and define in $S \cup 0$ a multiplication $*$ by: $x * 0 = 0 * x = 0$ for all $x \in S \cup 0$ and the pairs of elements of S retain their former product. If S has a zero, set $S^0 = S$, otherwise let S^0 be the semigroup S with a zero adjoined (a slightly different meaning is attached to S^0 in Clifford and Preston [1]). We will usually denote the identity of a semigroup by one of the symbols $1, e, \epsilon$, and the zero by 0.

I.4.6 NOTATION. For sets A and B, let $A \backslash B = \{a \in A | a \notin B\}$ and let $|A|$ be the cardinality of A. If S is a semigroup with zero 0, let $S^* = S \backslash 0$ with S^* inheriting the structure of a partial operation defined for the pairs of elements whose product in S is nonzero. (Hence S^* is a semigroup if and only if S has no zero divisors.)

I.4.7 DEFINITION. A semigroup in which every element is a left (right) zero is a *left* (*right*) *zero semigroup*. A semigroup S with zero 0 in which the product of any two elements equals 0 is a *zero semigroup*. A semigroup S with zero 0 is 0-*simple* if $S^2 \neq 0$ and S has no nonzero proper ideals.

It is clear that any nonempty set S admits the structure of a left zero, a right zero, or a zero semigroup. For the first, we define $ab = a$ and for the second $ab = b$ for all $a, b \in S$. For the third, we fix an element, denoted by 0, and let $ab = 0$ for all $a, b \in S$. The last condition can also be written as $S^2 = 0$. These semigroups exhibit many extreme properties, and for many properties P, some variant of the statement "a semigroup S has property P if and only if S is a left (or a right) or a zero semigroup" holds. In a zero semigroup S, ideals of S evidently coincide with subsets of S containing the zero. Hence a semigroup with zero having no proper nonzero ideals is either 0-simple or has at most 2 elements.

I.4.8 DEFINITION. Elements a and b of a semigroup S *commute* if $ab = ba$. The set

$$\mathcal{C}(S) = \{a \in S \mid as = sa \quad \text{for all} \quad s \in S\}$$

is the *center* of S. A semigroup S in which any two elements commute (i.e., $\mathcal{C}(S) = S$) is *commutative*.

I.4.9 DEFINITION. An element a of a semigroup S is *idempotent* if $a^2 = a$. An *idempotent semigroup*, or shorter a *band*, is a semigroup in which all elements are idempotent. A commutative band is a *semilattice*.

I.4.10 DEFINITION. A *subgroup* G of a semigroup S is a subsemigroup of S which is also a group.

I.4.11 PROPOSITION. Let e be an idempotent of a semigroup S. Then

$$\begin{aligned} G_e &= \{a \in S \mid a = ea = ae,\ e = aa' = a'a \quad \text{for some } a' \in S\} \\ &= \{a \in S \mid a \in eS \cap Se,\ e \in aS \cap Sa\} \end{aligned}$$

is the greatest subgroup of S having e as its identity.

Proof. It is clear that every subgroup of S having e as its identity is contained in the first set, and that the first set is contained in the second. A simple calculation shows that the first set is a subgroup of S having e as its identity. Let a be an element of the second set. Then $a = ex = ye$ and $e = az = wa$ for some $x,y,z,w \in S$. The first equation shows that $a = ea = ae$. Further $e = a(eze) = (ewe)a$ which implies

$$eze = e(eze) = e(wa)ze = ew(az)e = ewe$$

so that a is contained in the first set.

I.4.12 DEFINITION. If S is a semigroup with identity e, then G_e is the *group of units* of S, and its elements are the *invertible* elements of S.

I.4.13 LEMMA. An element a of a semigroup S with identity is invertible if and only if $aS = Sa = S$.

Proof. Exercise.

A generalization of the concepts of an idempotent element or semigroup is provided by the following notions.

I.4.14 DEFINITION. An element a of a semigroup S is *regular* if $a = axa$ for some $x \in S$. A semigroup S is *regular* if every element of S is regular.

I.4.15 DEFINITION. Let a be an element of a semigroup S. An element x of S is an *inverse* of a if $a = axa$, $x = xax$.

If in a semigroup we have $a = axa$, it is easy to verify that xax is an inverse of a. Hence every regular element has an inverse and conversely. For example, a group is a semigroup in which every element has a unique inverse.

I.4.16 DEFINITION. The *order* of a semigroup S is the number of its elements if S is finite, otherwise S is of *infinite order*. A semigroup of order 1 is a *trivial semigroup*. The *order of an element* s of a semigroup S is the order of the cyclic subsemigroup of S generated by s. A semigroup all of whose elements are of finite order is *periodic*.

I.5. Relations and Functions on a Semigroup

We will discuss here a few special kinds of binary relations and functions occurring most frequently in the study of semigroups.

I.5.1 DEFINITION. A *binary relation* ρ on a set A is a subset of the Cartesian product $A \times A$. We will write $a \rho b$ and say that a and b are *ρ-related* if $(a,b) \in \rho$ and will call ρ simply a *relation*. A relation ρ on A is

reflexive if $a \rho a$,
symmetric if $a \rho b$ implies $b \rho a$,
antisymmetric if $a \rho b$ and $b \rho a$ imply $a = b$,
transitive if $a \rho b$ and $b \rho c$ imply $a \rho c$

for all $a,b,c \in A$.

I.5.2 DEFINITION. A reflexive, symmetric, transitive relation ρ is an *equivalence relation*; its classes are *ρ-classes* and the ρ-class containing an element a will be denoted by $a\rho$. The relation ρ on A for which $a \rho b$ if and only if $a = b$ is the *equality relation* on A and will be denoted by ϵ_A; the relation ρ on A for which $a \rho b$ for all $a,b \in A$ is the *universal relation* on A and will be denoted by ω_A. Both ϵ_A and ω_A are equivalence relations; an equivalence relation on A is *proper* if it is different from ϵ_A.

I.5.3 DEFINITION. A reflexive, antisymmetric, transitive relation on a set P is a *partial order*; for it we often write $a \leq b$ and say that $(P,\leq)$, or simply P, is a *partially ordered set*. If B is a subset of a partially ordered set P, then an element a of P is an *upper bound* for B if $b \leq a$ for all $b \in B$; a *lower bound* is defined dually (in the context of partially ordered sets, "dually" means interchange $\leq$ and $\geq$, upper and lower, etc.). Further, a is a *greatest lower bound* of B if a is a lower bound of B and an upper bound of the set of all lower bounds of B. A *least upper bound* is defined dually. A partially ordered set P in which any two elements have a greatest lower bound is a *lower semilattice*; an *upper semilattice* is defined dually; if P is both an upper and a lower semilattice, it is called a *lattice*. A partial order $\leq$ on a set P is *linear* if for any $a,b \in P$, either $a \leq b$ or $b \leq a$; P is then said to be a *chain*. An element a of a partially ordered set P is the *least element* of P if $a \leq p$ for all $p \in P$, a is a *minimal element* of P if for any $x \in P$, $x \leq a$ implies $x = a$; the *greatest* and a *maximal element* of P are defined dually.

I.5.4 NOTATION. For any semigroup S, let E_S denote the set of all idempotents of S together with the binary relation defined by

$$e \leq f \quad \text{if (and only if)} \quad e = ef = fe$$

I.5.5 LEMMA. For any semigroup S, E_S is a partially ordered set.

Proof. Exercise.

So far we have encountered both a lower (and upper) semilattice as a special kind of partially ordered set, and a semilattice as a commutative idempotent semigroup. From this terminology we might reasonably expect a certain relationship among these notions.

I.5.6 PROPOSITION. Let S be a semilattice. Then E_S (equal to S as a set) is a lower semilattice with the greatest lower bound of $\{a,b\}$ equal to ab. Conversely, if T is a lower semilattice, then defining a multiplication on T by $a * b =$ greatest lower bound of $\{a,b\}$, T becomes a semilattice.

Proof. Exercise.

In light of this proposition, there is very often no need for distinguishing between a lower semilattice (a partially ordered set) and a semilattice (a semigroup). A usual example of such a semigroup is a linearly ordered set with multiplication defined by $a * b = \min \{a,b\}$, or using the opposite order, $a * b = \max \{a,b\}$. For the linearly ordered set, we may take, for example, the (positive, or nonnegative) integers or rationals or reals under the usual order.

I.5.7 DEFINITION. An equivalence relation ρ on a semigroup S is a *left congruence* if for all $a,b,c \in S$, $a \rho b$ implies $ca \rho cb$, a *right congruence* if $a \rho b$ implies $ac \rho bc$; ρ is a *congruence* if it is both a left and a right congruence. A *proper* (*left* or *right*) *congruence* is a (left or right) congruence which is proper as an equivalence relation.

I.5.8 LEMMA. An equivalence relation ρ on a semigroup S is a congruence if and only if for all $a,b,c,d \in S$, $a \rho b$ and $c \rho d$ imply $ac \rho bd$.

Proof. Exercise.

In view of this lemma, we are able to introduce the following concept.

I.5.9 DEFINITION. Let ρ be a congruence on a semigroup S. Then the set S/ρ of all ρ-classes with the multiplication $(a\rho)(b\rho) = (ab)\rho$ is the *quotient semigroup* relative to the congruence ρ.

The set of all binary relations on a set A inherits the Boolean operations of intersection, union, etc. as the set of subsets of $A \times A$. It is clear that the intersection of any set of congruences is again a congruence. We thus may introduce the following concept.

I.5.10 DEFINITION. The intersection of all congruences on a semigroup S containing a binary relation ρ on S is the *congruence generated* by ρ.

There are further a number of special kinds of functions on a semigroup which, like the left or right congruences, are in a certain way connected with the multiplication.

I.5.11 DEFINITION. Let S and T be semigroups. A function φ mapping S into T is a *homomorphism* of S into T if for all $a,b \in S$, we have $(a\varphi)(b\varphi) = (ab)\varphi$. If φ is one-to-one, then φ is an *isomorphism* or *embedding* of S into T, and S is said to be *embeddable* in T. If there is a homomorphism of S onto T, T is a *homomorphic image* of S; further, S and T are *isomorphic* if there is an isomorphism of S onto T; if so, we write $S \cong T$. A homomorphism of S into itself is an *endomorphism;* a one-to-one endomorphism of S onto itself is an *automorphism.*

I.5.12 LEMMA. If φ is a homomorphism of a semigroup S into a semigroup T, then the relation ρ on S defined by $a \rho b$ if and only if $a\varphi = b\varphi$, is a congruence on S, and $S/\rho \cong S\varphi$. Conversely, if ρ is a congruence on S, then the mapping $a \to a\rho$ is a homomorphism of S onto S/ρ.

Proof. Exercise.

I.5.13 DEFINITION. For a given homomorphism φ the congruence ρ defined in 5.12 is the congruence *induced* by φ. For a given congruence ρ, the mapping $a \to a\rho$ is the *natural homomorphism* of S onto S/ρ.

I.5.14 LEMMA. Let ρ be a congruence on a semigroup S. For every congruence γ on S containing ρ, define a binary relation γ' on S/ρ by

$$x\rho \, \gamma' \, y\rho \quad \text{if} \quad x \, \gamma \, y \qquad (x,y \in S).$$

Then the mapping $\gamma \to \gamma'$ is a one-to-one order preserving mapping of the set of all congruences on S containing ρ onto the set of all congruences on S/ρ.

Proof. Exercise.

I.5.15 NOTATION. For a nonempty set X, the notation

$$\varphi \colon x \to \bar{x} \qquad (x \in X)$$

means that φ is a function which maps x onto $\bar{x}$ for every $x \in X$. The identity function on X is denoted by ι_X and is written either on the left or the right, i.e., $\iota_X x = x\iota_X = x$ for all $x \in X$. If φ is a function of a set A into a set B we write $\varphi : A \to B$, and if C is a nonempty subset of B, then

$$C\varphi^{-1} = \{a \in A \mid a\varphi \in C\}.$$

If $C = \{c\}$, we write $c\varphi^{-1}$ instead of $\{c\}\varphi^{-1}$. Except for the functions $\mathfrak{a}$ and $\mathfrak{b}$ in V.3 and $\mathfrak{r}$ in V.6, all functions are denoted by lower case Greek letters and are written on the left or the right of the argument.

If σ is a function or a relation on a set A, and B is a nonempty subset of A, then $\sigma|_B$ denotes the *restriction* of σ to B, i.e., $\sigma|_B = \sigma \cap B \times B$. If X is a nonempty set, $\mathcal{P}(X)$ denotes the partially ordered set of all *nonempty* subsets of X under inclusion. The empty set is denoted by $\varnothing$.

The *dual* of an expression involving either concepts or products of elements of a semigroup is the expression obtained by interchanging the adjectives "left" and "right" and replacing each product ab by ba. If a statement A implies a statement B, then the dual of A clearly implies the dual of B. Dual statements will be omitted, but if C stands for the dual of B, we will mention sometimes that "dually C holds."

I.6. Examples

We will discuss only a sample of the most frequently occurring examples of semigroups.

I.6.1 REAL NUMBERS. Here we have the semigroups of positive (or nonnegative) integers under either addition or multiplication, positive (or nonnegative) rationals (or reals) under addition, etc. Furthermore, a different operation can be given to certain sets of real numbers, e.g., the set of positive integers with any of the operations

$$a * b = \min\{a,b\}, \qquad \max\{a,b\}, \qquad \text{g.c.d.}\{a,b\}, \qquad \text{l.c.m.}\{a,b\}.$$

I.6.2 TRANSFORMATIONS. A *transformation* on a set X is a function mapping X into itself. The *set* $\mathfrak{T}(X)$ *of all transformations on* X written on the left of the argument is a semigroup under the composition of functions $(\alpha\beta)x = \alpha(\beta x)$ for all $x \in X$; the same set with functions written on the right of the argument with the composition of functions $x(\alpha\beta) = (x\alpha)\beta$ for all $x \in X$ is a semigroup denoted by $\mathfrak{T}'(X)$. For example, the *set* $\mathfrak{T}_0(X)$ *of all constants* is a left zero semigroup and an ideal of $\mathfrak{T}(X)$. Further ideals of $\mathfrak{T}(X)$ are obtained by taking all transformations on X which map X into a subset of X of cardinality less than a fixed nonzero cardinal number. If X is endowed with some structure, e.g., of a group, a ring, a vector space, etc., we may consider subsemigroups of all endomorphisms of X relative to this structure. In the case of a vector space, linear transformations of rank less than a fixed nonzero cardinal number provide examples of ideals of the semigroup of all linear transformations on the vector space.

I.6.3 FREE SEMIGROUPS. Let X be a nonempty set. A nonempty finite sequence $a_1, a_2, \ldots, a_n$, usually written by juxtaposition, $a_1a_2 \ldots a_n$, of elements of X is called a *word over the alphabet* X. The set $\bar{X}$ of all words with the operation of juxtaposition

$$(a_1a_2 \ldots a_m)(b_1b_2 \ldots b_n) = a_1a_2 \ldots a_m\, b_1b_2 \ldots b_n$$

is a semigroup called the *free semigroup on the set X*.

Let β be the intersection of all congruences ρ on $\bar{X}$ for which $x^2 \,\rho\, x$ for all $x \in \bar{X}$. Then the quotient semigroup $\bar{X}/\beta$ is the *free band* on the set X. One defines the *free commutative semigroup* by using congruences ρ for which $xy \,\rho\, yx$ for all $x,y \in \bar{X}$; the *free semilattice*, by using congruences ρ for which both $x^2 \,\rho\, x$ and $xy \,\rho\, yx$ for all $x,y \in \bar{X}$. These semigroups indeed form a band, a commutative semigroup, and a semilattice, respectively, because the sets of these congruences are closed under intersection. The adjective "free" is associated with a property of universality which will not be discussed here.

I.6.4 MULTIPLICATIVE SEMIGROUP OF A RING. In view of the strongly developed theory of rings, this is a source of numerous examples of semigroups; e.g., semigroups of linear transformations on a vector space, mentioned in 6.2, and more generally, semigroups of endomorphisms of modules over arbitrary or restricted classes of rings. Further examples include the multiplicative semigroups of residue class rings of integers and various semigroups of matrices.

I.7. Exercises

1. Show that every infinite cyclic semigroup is isomorphic to the semigroup of positive integers under addition.

2. Give an example of a semigroup which does not have a finite generating set, and an example of a semigroup with a finite generating set but which is not cyclic.

3. Show that a subsemigroup of a cyclic semigroup need not be cyclic.

4. Find all ideals of and all congruences on the semigroup S defined on positive integers with the operation $x * y = \max \{x,y\}$.

5. On the set T of all nonempty subsets of a semigroup S define a multiplication by $A * B = AB$. Then T is a semigroup called the *power semigroup* of S. Find properties of semigroups which are preserved and properties which are not preserved by the passage from a semigroup to its power semigroup.

6. Show that a semigroup S is a group if and only if S is both left and right simple. Give an example of a left simple semigroup which is not a group.

7. Show that the left ideal generated by a nonempty subset A of a semigroup S equals $A \cup SA = S^1 A$ and that the ideal generated by A equals $A \cup SA \cup AS \cup SAS = S^1 A S^1$.

8.* For a finite cyclic semigroup S generated by a, let m be the smallest positive integer for which $a^m = a^n$ for some $n < m$. Prove that $\{a^n, a^{n+1}, \ldots, a^{m-1}\}$ is the unique maximal subgroup of S. What is the kernel of S? Deduce that S is a group if and only if $n = 1$. Also deduce that every finite semigroup contains an idempotent. (*Hint*: For the first statement consider the element a^{rd} where $r = m - n$, $n \leq d \leq m - 1$, and r and d are relatively prime.)

9. Let $S = \{x \mid 1 \leq x \leq 2 \quad \text{or} \quad 3 \leq x \leq 4 \quad \text{or} \quad x \geq 5\}$ with the operation

$$x * y = \begin{cases} x + y & \text{if} \quad x,y \geq 5, \\ \max \{x,y\} & \text{otherwise}. \end{cases}$$

Show that S is a semigroup and find all of its principal ideals. Does S have a kernel? What are the orders of elements of S?

10. Find all left and all right ideals of a (i) zero semigroup, (ii) left zero semigroup.

11. Let S be the set of all 2×2 matrices with one row of zeros and the other row with coefficients in the interval $[-1,1]$. Show that S is a semigroup under multiplication and find all of its idempotents.

12. Let S be a nonempty set and 0 be a fixed element of S. On S define an operation by:

$$x * y = \begin{cases} x & \text{if } x = y, \\ 0 & \text{otherwise}. \end{cases}$$

Show that S is a semigroup (called a *Kronecker semigroup*) and find all of its idempotents and ideals.

13. Let $S = \{1,2,3, \ldots\} \cup \{1',2',3', \ldots\}$ with the operation

$$x * y = x' * y' = \max\{x,y\}, \qquad x' * y = x * y' = (\max\{x,y\})'.$$

Show that S is a semigroup and find all of its idempotents and principal ideals. Does S have a kernel?

14.* Prove that the following conditions on a semigroup S are equivalent.
   i) For every $a \in S$ there exists a unique $x \in S$ such that $ax \in E_S$.
   ii) For every $a \in S$ there exists a unique $x \in S$ such that $a = axa$.
   iii) S is a regular semigroup containing exactly one idempotent.
   iv) S is a group.
   (*Hint:* Prove the implications in circular order.)

15. Show that every finite semigroup is periodic and that the converse does not hold.

16. Show that in the following semigroups every equivalence relation is a congruence: (i) zero semigroups; (ii) semilattices of order 2; (iii) left zero semigroups; (iv) right zero semigroups.

17.* Prove that a semigroup with zero in which every equivalence relation is a congruence is either of type (i) or (ii) in the preceding exercise.

18. Find the semigroup of endomorphisms of an infinite cyclic semigroup.

19. Show that the set of all congruences on a semigroup S containing a fixed congruence on S is a lattice under inclusion.

20. In the semigroup $\mathfrak{I}(X)$, characterize all elements α with the property $\alpha\mathfrak{I}(X) = \mathfrak{I}(X)$, and those satisfying $\mathfrak{I}(X)\alpha = \mathfrak{I}(X)$.

21.* Find all idempotents and the respective maximal subgroups of $\mathfrak{I}(X)$. Does $\mathfrak{I}(X)$ have left zeros, right zeros, a kernel? Determine the partial order in $E_{\mathfrak{I}(X)}$. Prove that $\mathfrak{I}(X)$ is a regular semigroup.

22. Give an example of a semigroup of transformations without idempotents.

23. Find the order of elements of a free semigroup S on a set. Does S have a kernel?

24. Give an example of a 0-simple semigroup. (*Hint:* Consider semigroups of $n \times n$ matrices over a field.)

25.* Prove that every semigroup is isomorphic to a subsemigroup of the multiplicative semigroup of some ring. Give an example of a semigroup which is not isomorphic to the multiplicative semigroup of any ring.

26. What is the group of units, the set of zero divisors and the kernel of the semigroup of all $n \times n$ matrices over a field?

I.8. REFERENCES

The general references on the theory of "algebraic" or "abstract" semigroups are Ljapin's book [2] and the two volumes of Clifford and Preston [1]. The book of Suškevič [2] is mainly of historical importance and is practically not available. Rédei's book [1] is of interest in the special topic it considers and is available both in German and in English. The little book of Papy [1] is intended as an introduction to "groupoids," actually semigroups, for high school students and has a very colorful character.

Several books contain a more or less extensive discussion of semigroups. Bruck's book [1] gives a review of binary systems and includes a survey of a part of semigroup theory (Chapter 2). Dubreil's text [2] contains an extensive discussion of what is sometimes referred to as the "French school of semigroups" (Chapter 5). The collection of problems [1] of Ljapin, Aizenštat and Lesohin contains many elementary problems on semigroups, particularly on semigroups of transformations (Chapters 1, 2, 3). Chevalley's monograph [1] contains a discussion of monoids (semigroups with an identity) (Chapter 1). Rosenfeld's textbook [1] includes a discussion of semigroups in general (Chapter 6). A. Ginsburg's book [1] has an introductory discussion of semigroups to be used in automata theory (Chapters 1, 7).

The books of Birkhoff [1] (Chapter 14), Dubreil-Jacotin, Lesieur and Croisot [1] (Part II) and Fuchs [1] (Part III) contain an extensive study of ordered semigroups. The principal references for topological semigroups are the monographs of Paalman-de Miranda [1] and Hofmann and Mostert [1].

Several lecture notes are devoted either fully or in part to the theory of semigroups. Kapp and Schneider [1] consists of a detailed discussion of the congruences on a completely 0-simple semigroup. Krohn, Rhodes and Tilson [1] in two volumes is a pioneering study of finite semigroups and is mainly designed as an introduction to automata theory (a large part of these notes appeared in Arbib [1]). The author's notes [9] cover many topics in the theory of abstract and transformation semigroups, with a large overlapping with this book; the notes [15] deal with the semigroups of linear transformations on a vector space. Schein [5] covers a topic of the theory of transformation semigroups.

A number of collections of articles dealing mainly or entirely with the subject of semigroups have appeared. In the English language there are collections of papers edited by Saitô [1], Arbib [1] and Folley [1]; in Russian by Vagner [1]. Furthermore there exists a profusion of reports on general algebra or the semigroup theory at all-union or regional conferences in the Soviet Union containing a great number of research announcements without proofs, e.g., the conferences in Tartu 1966, Riga 1967 etc.

A few review articles can be helpful in gaining a global picture of certain topics in semigroups as well as obtaining a complete list of the literature on the subject. In first place are two articles of Gluskin [6], [7] and an article of Gluskin, Schein and Ševrin [1] which appeared in *Itogi Nauki* and briefly cover all articles which were reviewed in the *Referativnii Žurnal* in the years 1960–1966. Williamson's article [1] covers many topics in the harmonic analysis on semigroups, the author's [16] discusses in detail ideal extensions of semigroups and briefly touches upon extensions of rings and partially ordered sets in a more or less unified way.

Several tables of semigroups are available. In chronological order they are: Tamura [1] up to order 4 computed by hand, Forsyth [1] of order 4 using a computer, Tamura *et al.* [1] up to order 5 computed by hand, Selfridge [1] of order 5, Plemmons [1] up to order 6, the last two using a computer. These tables are particularly useful in searching for a (small) counterexample or verifying a conjecture on semigroups of small order, which essentially amounts to the same thing.

The last but certainly not the least, the *Semigroup Forum*, a journal published by the Springer Verlag and entirely devoted to questions in the theory of semigroups — algebraic, topological, ordered etc. — publishes many survey articles, research announcements, brief notes, problems and bibliographical information, of definite interest to anyone concerned with this subject.

Semilattice Decompositions

The point of view adopted throughout this book is that we can study the structure of a relatively complex semigroup by first decomposing it into a collection of subsemigroups each of which has somewhat simpler structure and, conversely, composing a complicated and "bigger" semigroup from simpler ones. The decomposition and composition in question are of a very special kind; viz., the greatest semilattice decomposition and a semilattice composition, respectively. These have proved the most fruitful among various decompositions and compositions. In this chapter we will study only such decompositions, in considerable detail. This leads, on the one hand, to the study of subdirect products, completely prime ideals, filters and related subjects, and on the other, to the consideration of various classes of semigroups having some special properties with respect to their greatest semilattice decompositions.

II.1. Subdirect Products

This section is mainly of universal-algebraic character, and provides the frame work for the treatment of semilattice decompositions of a semigroup. Specifically, after the needed definitions have been set, we prove a few basic theorems concerning subdirect products and their connection with congruences on a semigroup. A prominent role is played here by the notions of subdirect irreducibility, $\mathfrak{C}$-congruences on a semigroup (i.e., congruences ρ relative to which S/ρ belongs to a certain class $\mathfrak{C}$ of semigroups), $\mathfrak{C}$-decompositions and the greatest ones among these. These concepts and their properties represent a minimum needed for the next section of this chapter and the material in IV.5.

II.1.1 DEFINITION. Let $\{S_\alpha\}_{\alpha \in A}$ be a nonempty family of semigroups. The semigroup S defined on the Cartesian product of the sets S_α with coordinatewise multiplication, i.e., $(x_\alpha)(y_\alpha) = (x_\alpha y_\alpha)$, is the *direct product* of the semigroups $\{S_\alpha\}_{\alpha \in A}$ and is denoted by $S = \Pi_{\alpha \in A} S_\alpha$. Any semigroup isomorphic to S is also called the direct product of the semigroups $\{S_\alpha\}_{\alpha \in A}$.

For convenience we will tacitly assume that no S_α is trivial. For $\beta \in A$, the function $\pi_\beta : S \to S_\beta$ defined by $(x_\alpha)\pi_\beta = x_\beta$ for all $(x_\alpha) \in S$ is the *projection homomorphism* of S onto the *β-component* S_β.

The direct product is a very restrictive construction in the sense that a semigroup rarely can be represented as a direct product of sufficiently simple semigroups. This situation is sometimes remedied by introducing the following generalization.

II.1.2 DEFINITION. A semigroup S is a *subdirect product* of semigroups $\{S_\alpha\}_{\alpha \in A}$ if S is isomorphic to a subsemigroup T of $\Pi_{\alpha \in A} S_\alpha$ which has the property that $T\pi_\alpha = S_\alpha$ for all $\alpha \in A$. We may assume that no S_α is trivial and identify S with T. A nontrivial semigroup S is *subdirectly irreducible* if it has the property: When S is a subdirect product of semigroups $\{S_\alpha\}_{\alpha \in A}$, then for at least one $\beta \in A$, π_β maps S onto S_β isomorphically.

We should point out at this stage that the description of a semigroup as a subdirect product of certain semigroups is vague; indeed, there may be a great many subdirect products of the same family of semigroups and these may be very difficult to find.

II.1.3 DEFINITION. A congruence σ on a semigroup S *separates* the elements x and y of S if x and y are contained in different σ-classes. A family Σ of congruences on S *separates elements* of S if for every pair x and y of distinct elements of S, there exists $\sigma \in \Sigma$ which separates x and y.

It is clear that Σ separates elements of S if and only if $\bigcap_{\sigma \in \Sigma} \sigma = \epsilon_S$.

II.1.4 PROPOSITION. If a semigroup S is a subdirect product of semigroups $\{S_\alpha\}_{\alpha\in A}$, then the set of congruences $\{\sigma_\alpha\}_{\alpha\in A}$ on S induced by the different π_α separates elements of S. Conversely, if $\{\sigma_\alpha\}_{\alpha\in A}$ is a family of congruences on S, all different from the universal relation, which separates elements of S, then S is a subdirect product of the semigroups $\{S/\sigma_\alpha\}_{\alpha\in A}$.

Proof. Let S, S_α and σ_α be as in the first part of the proposition. If $x,y \in S$, $x \neq y$, then writing $x = (x_\alpha)$, $y = (y_\alpha)$, there exists $\beta \in A$ such that $x_\beta \neq y_\beta$. But then $x\ \sigma_\beta\ y$ does not hold. Hence the family of congruences $\{\sigma_\alpha\}_{\alpha\in A}$ separates elements of S.

Conversely, let $\{\sigma_\alpha\}_{\alpha\in A}$ be a family of congruences on S such that $\bigcap_{\alpha\in A}\sigma_\alpha = \epsilon_S$. Define a function $\varphi:S \to \Pi_{\alpha\in A}S/\sigma_\alpha$ by:

$$\varphi:x \to (x\sigma_\alpha) \qquad (x \in S).$$

It is easy to verify that φ is a homomorphism and that $S\varphi\pi_\alpha = S/\sigma_\alpha$ for every $\alpha \in A$. If x and y are distinct elements of S, then by hypothesis σ_β exists such that $x\ \sigma_\beta\ y$ does not hold, so that $x\sigma_\beta \neq y\sigma_\beta$. But then $x\varphi \neq y\varphi$, which proves that φ is one-to-one. Thus S is a subdirect product of the semigroups $\{S/\sigma_\alpha\}_{\alpha\in A}$.

II.1.5 COROLLARY. The following conditions on a nontrivial semigroup S are equivalent.

i) S is subdirectly irreducible.
ii) The intersection of any set of proper congruences on S is a proper congruence on S.
iii) S has a least proper congruence.

Proof. Exercise.

II.1.6 COROLLARY. A semigroup S is a subdirect product of semigroups $\{S_\alpha\}_{\alpha\in A}$ if and only if for every $\alpha \in A$ there is a homomorphism of S onto S_α and the family of induced congruences separates elements of S.

Proof. Exercise.

II.1.7 COROLLARY. If S is a subdirect product of semigroups $\{S_\alpha\}_{\alpha\in A}$ and for each $\alpha \in A$, S_α is a subdirect product of semigroups $\{S_{\alpha,\beta}\}_{\beta\in A_\alpha}$, then S is a subdirect product of semigroups $S_{\alpha,\beta}$.

Proof. Exercise.

Thus we can say, roughly, that the property of being a subdirect product is transitive. The next theorem is a specialized version of a theorem valid for universal algebras; the proof for universal algebras is essentially the same.

II.1.8 THEOREM. Every semigroup is a subdirect product of subdirectly irreducible semigroups.

Proof. Let S be a semigroup. For every pair x,y of different elements of S, let $M(x,y)$ denote the set of all congruences on S which separate x and y. Then $\epsilon_S \in M(x,y)$ so that $M(x,y) \neq \varnothing$.

We fix a pair $x,y \in S$ with $x \neq y$, and wish to show that the partially ordered set (under inclusion) $M(x,y)$ has a maximal element. By Zorn's lemma it suffices to show that every chain in $M(x,y)$ has an upper bound. Hence let Γ be a chain in $M(x,y)$ and define a relation τ on S by: $z \, \tau \, w$ if there exists $\sigma \in \Gamma$ such that $z \, \sigma \, w$. Then τ is evidently reflexive and symmetric; if $z \, \tau \, w$ and $w \, \tau \, v$, then $z \, \sigma \, w$ and $w \, \sigma' \, u$ for some $\sigma,\sigma' \in \Gamma$. If $\sigma \subseteq \sigma'$, then also $z \, \sigma' \, w$ and hence $z \, \sigma' \, u$ so that $z \, \tau \, u$; the case $\sigma' \subseteq \sigma$ is symmetric. Thus τ is transitive and is an equivalence relation. If $z \, \tau \, w$, then $z \, \sigma \, w$ for some $\sigma \in \Gamma$, so that $zv \, \sigma \, wv$ for every $v \in S$, and we have that $zv \, \tau \, wv$; similarly $vz \, \tau \, vw$. Thus τ is a congruence. Also τ separates x and y since each σ separates x and y. Consequently, $\tau \in M(x,y)$ and is evidently an upper bound of Γ.

For every pair x,y of different elements of S, choose a maximal congruence $\sigma(x,y)$ separating x and y. Then the family of congruences $\sigma(x,y)$ separates elements of S, and hence by 1.4, S is a subdirect product of the semigroups $S/\sigma(x,y)$.

According to and in the notation of I.5.14, the intersection γ of all congruences on S properly containing $\sigma(x,y)$ corresponds to the intersection γ' of all proper congruences on $S/\sigma(x,y)$. By the maximality of $\sigma(x,y)$, γ does not separate x and y, which implies that $\gamma \neq \sigma(x,y)$. But then $\gamma' \neq \epsilon_{S/\sigma(x,y)}$ which proves that $S/\sigma(x,y)$ is subdirectly irreducible by 1.5.

Note that if a semigroup S is a subdirect product of semigroups $\{S_\alpha\}_{\alpha \in A}$, then each S_α is a homomorphic image of S. Hence 1.8 remains valid if in it we substitute "the class of all semigroups" by any class of semigroups closed under taking homomorphic images. For example, it follows that every (idempotent, commutative, or both) semigroup is a subdirect product of subdirectly irreducible (idempotent, commutative, or both) semigroups. For it is clear that the classes of all (idempotent, commutative, or both) semigroups are closed under taking homomorphic images. One usually requires that a class of semigroups not only be closed under taking homomorphic images, but also under taking subsemigroups and direct products; e.g., the classes of all semigroups, or of all bands, or of all semilattices have this property. To insure this, one usually introduces the following universal algebraic concept.

II.1.9 DEFINITION. A class $\mathcal{C}$ is *equationally defined*, or is a *variety*, if there exists a set E of equations such that a semigroup S is in $\mathcal{C}$ if and only if S satisfies all of the equations in E identically.

The last phrase means that for every substitution of variables in an equation by elements of the semigroup, the resulting elements on each side of the equation are equal.

For example, for $E = \{x = x\}$, we get the class of all semigroups; for $E = \{xy = yx\}$, the class of all commutative semigroups; for $E = \{x^2 = x, xy = yx\}$, the class of all semilattices, etc. It is evident that a variety is closed under taking of homomorphic images, subsemigroups, and direct products.

II.1.10 DEFINITION. If $\mathcal{C}$ is any class of semigroups, S is a semigroup and σ is a congruence on S, then σ is a $\mathcal{C}$-*congruence* if $S/\sigma \in \mathcal{C}$. If $\mathcal{C}$ is the class of all semilattices, $\mathcal{C}$-congruences are called *semilattice congruences*; one defines analogously *band congruences*, *left zero congruences*, etc.

For example, a congruence σ on a semigroup S is a semilattice congruence if and only if for all $x,y \in S$, $xy \,\sigma\, yx$, $x^2 \,\sigma\, x$. Similar expressions hold for other congruences.

II.1.11 THEOREM. Let $\mathcal{C}$ be a variety of semigroups, $\mathcal{D}$ the class of all subdirectly irreducible semigroups in $\mathcal{C}$, S any semigroup. Then a congruence σ on S, different from the universal congruence, is a $\mathcal{C}$-congruence if and only if σ is the intersection of $\mathcal{D}$-congruences.

Proof. *Necessity*. Let σ be a $\mathcal{C}$-congruence on S different from the universal congruence, and let $\varphi: S \to T = S/\sigma$ be the natural homomorphism. Then $T \in \mathfrak{C}$ and is nontrivial. According to 1.8, T is a subdirect product of semigroups $\{S_\alpha\}_{\alpha \in A}$ where each S_α is in $\mathfrak{D}$. By 1.6, for every $\alpha \in A$, there is an onto homomorphism $\varphi_\alpha : T \to S_\alpha$. Consequently, for every $\alpha \in A$, $\varphi\varphi_\alpha$ maps S homomorphically onto S_α; let σ_α be the induced congruence. If $x \,\sigma\, y$, then $x\varphi = y\varphi$ so for every $\alpha \in A$, we have $x\varphi\varphi_\alpha = y\varphi\varphi_\alpha$ and thus $x \,\sigma_\alpha\, y$. Consequently $x \bigcap_{\alpha \in A} \sigma_\alpha\, y$ and thus $\sigma \subseteq \bigcap_{\alpha \in A} \sigma_\alpha$. Conversely if $x \bigcap_{\alpha \in A} \sigma_\alpha\, y$, then for every $\alpha \in A$ we have $x \,\sigma_\alpha\, y$, whence also $(x\varphi)\varphi_\alpha = (y\varphi)\varphi_\alpha$. Since T is a subdirect product of semigroups S_α, by 1.6 the family of congruences on T induced by the homomorphisms φ_α, separates elements of T, which implies that $x\varphi = y\varphi$. But then $x \,\sigma\, y$, and hence $\bigcap_{\alpha \in A} \sigma_\alpha \subseteq \sigma$. Therefore $\sigma = \bigcap_{\alpha \in A} \sigma_\alpha$ where each σ_α is a $\mathfrak{D}$-congruence.

Sufficiency. Let $\sigma = \bigcap_{\alpha \in A} \sigma_\alpha$ where each σ_α is a $\mathfrak{D}$-congruence. Let $\varphi : S \to T = S/\sigma$ be the natural homomorphism, and in T, for every $\alpha \in A$, define a relation τ_α by: $x\varphi \,\tau_\alpha\, y\varphi$ if $x \,\sigma_\alpha\, y$. Then each τ_α is a congruence on T and the family of congruences $\{\tau_\alpha\}_{\alpha \in A}$ separates elements of T so that T is a subdirect product of semigroups T/τ_α and $T/\tau_\alpha \cong S/\sigma_\alpha$. Thus T is a subdirect product of semigroups in $\mathfrak{D}$. If E is a set of equations defining the class $\mathfrak{C}$, then every equation in E holds identically in each T/τ_α and thus also in T, which proves that σ is a $\mathfrak{C}$-congruence.

II.1.12 COROLLARY. With $\mathfrak{C}$ and $\mathfrak{D}$ as in 1.11, and S a nontrivial semigroup, we have $S \in \mathfrak{C}$ if and only if the family of $\mathfrak{D}$-congruences on S separates elements of S.

Let $\mathfrak{C}$ and $\mathfrak{D}$ be as in 1.11. The main use of 1.11 is that we obtain all $\mathfrak{C}$-congruences by taking arbitrary intersections of different families of $\mathfrak{D}$-congruences. This can be done effectively only when we are able to construct all $\mathfrak{D}$-congruences, which in turn requires the knowledge of the class $\mathfrak{D}$. It turns out that for several classes $\mathfrak{C}$ of semigroups (or other algebraic systems), the corresponding class $\mathfrak{D}$ is very small (for example, consists (up to isomorphism) of very few elements), or is infinite but can be explicitly given. If $\mathfrak{D}$ is known, one can sometimes construct all $\mathfrak{D}$-congruences, whose arbitrary intersections then give all $\mathfrak{C}$-congruences. We will do this for the class of all semilattices. In particular, the intersection μ of all $\mathfrak{C}$-congruences (or, what is the same, the intersection of all $\mathfrak{D}$-congruences) on a semigroup S is the *least* $\mathcal{C}$-*congruence* on S; μ has the following universal property: The diagram

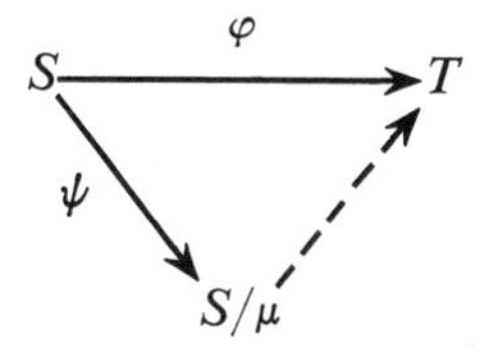

can be uniquely completed to a commutative diagram whenever φ is a homomorphism of S onto a semigroup T in $\mathcal{C}$ (here ψ is the natural homomorphism); also $\mu = \epsilon_S$ if and only if $S \in \mathcal{C}$.

II.1.13 DEFINITION. The partition of a semigroup S induced by a $\mathcal{C}$-congruence is a $\mathcal{C}$-*decomposition* of S. The partition of S induced by the least $\mathcal{C}$-congruence is the *greatest* $\mathcal{C}$-*decomposition* of S; and the corresponding quotient semigroup, the *greatest* $\mathcal{C}$-*homomorphic image* of S.

This terminology may seem confusing, but if one observes that the partially ordered set of congruences is ordered by inclusion of binary relations, while a decomposition is "greater," i.e., has a "greater" number of classes when the corresponding congruence is "smaller," this terminology should cause no difficulties. A $\mathcal{C}$-decomposition can be defined for any class $\mathcal{C}$ of semigroups, but we will be interested here only in varieties (otherwise the greatest $\mathcal{C}$-decomposition need not exist), in particular when $\mathcal{C}$ is the class of semilattices. The property of S/μ relative to being "greatest" can be interpreted conveniently using the diagram above, for one shows easily that any $\mathcal{C}$-homomorphic image of S is a homomorphic image of S/μ.

The concept in the following definition will be used extensively.

II.1.14 DEFINITION. A semigroup S is a *semilattice of semigroups belonging to a class* $\mathcal{C}$ if there exists a semilattice congruence σ on S all of whose classes belong to $\mathcal{C}$. The concepts: a *band* or a *left zero semigroup of semigroups belonging to* $\mathcal{C}$, are defined analogously.

The class $\mathcal{C}$ is usually taken to be quite restricted, e.g., the class of groups, the class of simple semigroups, etc.

II.1.15. Exercises

1. Give an example of a semigroup which has a minimal but no least proper congruence.

2. Show that the two systems of equations $E = \{x = xyx\}$ and $E' = \{x^2 = x, xy = xzy\}$ define the same variety of semigroups.

3. Prove that a group G is subdirectly irreducible if and only if G has a nontrivial normal subgroup contained in every nontrivial normal subgroup of G. Give examples of such groups.

4.* Find all subdirectly irreducible cyclic semigroups. (*Hint:* See I.7.8 and show that an infinite cyclic semigroup is a subdirect product of finite cyclic semigroups.)

5. Let σ and τ be congruences on a semigroup S, and define a mapping $\chi: s \longrightarrow (s\sigma, s\tau)$. Show that χ is a homomorphism mapping S onto a subdirect product of S/σ and S/τ. Find necessary and sufficient conditions on σ and τ in order for χ to be (i) one-to-one, (ii) onto $S/\sigma \times S/\tau$, (iii) both.

6.* Let S be a semigroup and R be a nontrivial right zero semigroup. Prove that $S \times R$ is the only subdirect product of S and R contained in $S \times R$ if and only if S is left simple.

II.1.16 REFERENCES: Schein [1], [4], Tamura [11], Tamura and Kimura [2], Thierrin [6], Yamada [4].

II.2. Completely Prime Ideals and Filters

We now approach the main subject of this chapter, the characterization of semilattice congruences by means of completely prime ideals. We will follow the procedure outlined in the preceding section, i.e., we will first find all subdirectly irreducible semilattices, then construct all congruences induced by homomorphisms onto these, and finally take arbitrary intersections of the congruences constructed.

II.2.1 PROPOSITION. The two-element chain is, up to isomorphism, the only subdirectly irreducible semilattice.

Proof. Let S be a semilattice. For every $a \in S$, let σ_a be the relation on S defined by

$$x \; \sigma_a \; y \quad \text{if either} \quad x,y \geq a \quad \text{or} \quad x,y \ngeq a.$$

Then σ_a is a congruence on S having at most two classes. It then follows that $\sigma_a = \epsilon_S$ for some $a \in S$ if and only if $|S| \leq 2$. Further, $\bigcap_{a \in S} \sigma_a = \epsilon_S$ since $\sigma_a = \sigma_b$ implies that $a = b$. If $|S| > 2$, then for all $a \in S$, σ_a is a proper congruence, so that 1.5 implies that S is subdirectly reducible.

II.2.2 DEFINITION. An ideal I of a semigroup S is *completely prime* if for any $a,b \in S$, $ab \in I$ implies that either $a \in I$ or $b \in I$.

Note that an ideal I of S is completely prime if and only if $S \backslash I$ is either a subsemigroup of S or is empty. (The concept of a prime ideal will be defined at the beginning of the next section.)

II.2.3 PROPOSITION. Let $Y = \{0,1\}$ be the two-element chain of integers 0 and 1. Then a function φ mapping a semigroup S onto Y is a homomorphism if and only if $0\varphi^{-1}$ is a completely prime ideal of S.

Proof. Exercise.

Combining these two propositions with the results of the preceding section, we see that every semilattice congruence on a semigroup S is the intersection of congruences σ_I where I is a completely prime ideal and for all $x,y \in S$, $x \; \sigma_I \; y$ if and only if $x,y \in I$ or $x,y \notin I$.

II.2.4 NOTATION. Let S be a semigroup. Let $\mathcal{I}_S$ be the set of all proper completely prime ideals of S together with the empty set, and let $\mathcal{P}(\mathcal{I}_S)$ be the set of all nonempty subsets of $\mathcal{I}_S$ partially ordered under inclusion. Also let $\mathcal{E}_S$ denote the partially ordered set of all semilattice congruences on S. (Note that $\mathcal{E}_S \neq \varnothing$ and that congruences are ordered as binary relations, i.e., as subsets of $S \times S$ under inclusion.) For $\mathcal{A} \in \mathcal{P}(\mathcal{I}_S)$, let $\sigma_{\mathcal{A}}$ be the binary relation on S defined by $x \; \sigma_{\mathcal{A}} \; y$ if for every $I \in \mathcal{A}$, either $x,y \in I$ or $x,y \notin I$.

The next theorem is of basic importance for a large part of this chapter.

II.2.5 THEOREM. The function $\varphi : \mathcal{A} \rightarrow \sigma_{\mathcal{A}}$ maps $\mathcal{P}(\mathcal{I}_S)$ onto $\mathcal{E}_S$ and is inclusion inverting.

Proof. In view of the above discussion, it suffices to remark that for $\mathfrak{A} \in \mathcal{P}(\mathcal{I}_S)$, we have $\sigma_{\mathfrak{A}} = \bigcap_{I \in \mathfrak{A}} \sigma_I$, and that $\mathfrak{A} \subseteq \mathfrak{B}$ with $\mathfrak{A},\mathfrak{B} \in \mathcal{P}(\mathcal{I}_S)$ implies $\bigcap_{I \in \mathfrak{B}} \sigma_I \subseteq \bigcap_{I \in \mathfrak{A}} \sigma_I$ so that φ is inclusion inverting.

The inclusion of $\varnothing$ in $\mathcal{I}_S$ and the exclusion of S is needed in order to obtain an inclusion inverting function. The next corollary sets the notation to be used throughout this book.

II.2.6 COROLLARY. The congruence $\mathfrak{N} = \sigma_{\mathcal{I}_S}$ is the least semilattice congruence on S; hence $Y_S = S/\mathfrak{N}$ is the greatest semilattice homomorphic image of S, and the set of $\mathfrak{N}$-classes of S constitutes the greatest semilattice decomposition of S.

We will next characterize the $\mathfrak{N}$-classes more precisely. The following concept will come in handy.

II.2.7 DEFINITION. A subsemigroup F of a semigroup S is a *filter* of S if for all $x,y \in S$, $xy \in F$ implies $x,y \in F$.

It is easy to verify that a nonempty subset F of S is a filter if and only if $S \backslash F$ is either empty or is a completely prime ideal. Thus the filters of S are precisely the complements of elements of $\mathcal{I}_S$. Also, the nonempty intersection of filters is easily seen to be a filter. Hence, for every $x \in S$, the intersection of all filters containing x is again a filter, and thus the least filter containing x.

II.2.8 NOTATION. For any element x of a semigroup S, let $N(x)$ denote the *least filter* of S containing x, and let

$$N_x = \{y \in S \mid N(x) = N(y)\}.$$

II.2.9 PROPOSITION. Let S be a semigroup. For any $x \in S$, N_x is the $\mathfrak{N}$-class of S containing x. The elements of Y_S are precisely the different sets N_x. The multiplication in Y_S is given by $N_x N_y = N_{xy}$ and has the properties $N_{xy} = N_{yx}$, $N_{x^2} = N_x$ for all $x,y \in S$.

Proof. The first statement follows easily from the above remarks, the second is obvious, the third follows from the fact that $\mathfrak{N}$ is a congruence and the fourth is valid since Y_S is a semilattice.

Recall I.3.6 and I.3.7

II.2.10 PROPOSITION. Let x be an element of a semigroup S. Let $N_1(x)$ be the semigroup generated by x. Supposing that $N_n(x)$ has been defined for $n \geq 1$, let $N_{n+1}(x)$ be the semigroup generated by all elements y of S for which $N_n(x) \cap J(y) \neq \varnothing$. Then $N(x) = \bigcup_{n=1}^{\infty} N_n(x)$.

Proof. Let $T = \bigcup_{n=1}^{\infty} N_n(x)$. Clearly $N_n(x) \subseteq N_{n+1}(x)$ for $n = 1,2, \ldots$. Since each $N_n(x)$ is a semigroup, T is also a semigroup. Moreover, if $yz \in T$, then $yz \in N_n(x) \cap J(y) \cap J(z)$ for some n and thus $y,z \in N_{n+1}(x) \subseteq T$. Hence T is a filter and also $x \in N_1(x) \subseteq T$. If F is any filter of S containing x, then by induction, we see that each $N_n(x)$ is contained in F by the definition of a filter. Hence $T = N(x)$.

The next theorem concerns the structure of an $\mathfrak{N}$-class.

II.2.11 THEOREM. If I is an ideal of an $\mathfrak{N}$-class of a semigroup, then I has no proper completely prime ideals.

Proof. Let S be any semigroup, z be an element of S and I be an ideal of N_z. It suffices to show that I, itself, is the only filter of I. Hence let F be a filter of I, a be an element of F, and let

$$T = \{x \in S \mid a^2x \in F\}.$$

We show next that T is a filter of S.

Let $x,y \in T$; then $a^2y \in F$, which together with the inclusions $F \subseteq I \subseteq N_z$, implies $N_{ya} = N_{a^2y} = N_z$. Hence $ya \in N_z$ and thus $ya^2 \in I$. Further, $a^2(ya^2) = (a^2y)a^2 \in F$ so that $ya^2 \in F$. Similarly, $a^2x \in F$ implies $ax \in N_z$, which in turn yields $axy \in N_z$ since $N_{axy} = N_{ax}N_{ay} = N_z$. Consequently $a^2xy \in I$; on the other hand, $a^2x,ya^2 \in F$ implies $(a^2x)(ya^2) \in F$. But then also $(a^2xy)a^2 \in F$ so that $a^2xy \in F$. Therefore $xy \in T$.

Conversely, let $xy \in T$. Then $a^2xy \in F$, and thus also $(a^2x)\,(ya^2) = (a^2xy)a^2 \in F$. Moreover, $a^2xy \in F$ easily implies ax, $ya \in N_z$ and hence $a^2x,ya^2 \in I$. But since $(a^2x)(ya^2) \in F$, it follows that $a^2x,ya^2 \in F$. As before, we infer that $a^2y \in F$ since $(a^2y)a^2 = a^2(ya^2) \in F$ and $a^2y \in I$. Thus $x,y \in T$.

Hence T is a filter of S. It is easy to see that $T \cap I = F$. But then $a \in N_z \cap T$ implies that $N_z \subseteq T$ and therefore $T \cap I = I$, i.e., $F = I$.

In view of this theorem, it is convenient to introduce the following concept.

II.2.12 DEFINITION. A semigroup without proper completely prime ideals is $\mathfrak{N}$-*simple*.

The above theorem can be phrased thus: every ideal of an $\mathfrak{N}$-class is $\mathfrak{N}$-simple. Furthermore, a semigroup is $\mathfrak{N}$-simple if and only if it is semilattice indecomposable (i.e., has only a trivial semilattice homomorphic image).

II.2.13 COROLLARY. Every semigroup is a semilattice of $\mathfrak{N}$-simple semigroups.

II.2.14 COROLLARY. Every completely prime ideal of a semigroup S is a union of $\mathfrak{N}$-classes.

Proof. If I is a completely prime ideal and $I \cap N_x \neq \emptyset$, then $I \cap N_x$ is a completely prime ideal of N_x and by 2.11, we must have $I \cap N_x = N_x$ and thus $N_x \subseteq I$.

II.2.15 COROLLARY. If I is a completely prime ideal of a semigroup S, then $J = \{N_x \in Y_S | \ x \in I\}$ is a completely prime ideal of Y_S. Conversely, if J is a completely prime ideal of Y_S, then $I = \{x \in S | \ N_x \in J\}$ is a completely prime ideal of S. This establishes a one-to-one, order preserving (relative to inclusion) correspondence between the partially ordered set of all completely prime ideals of S and the partially ordered set of all completely prime ideals of Y_S.

Proof. Exercise.

II.2.16 PROPOSITION. Every $\mathfrak{N}$-simple subsemigroup T of a semigroup S is contained in an $\mathfrak{N}$-class of S.

Proof. If $T \nsubseteq N_x$ for all $x \in S$, then there exist $x,y \in T$ such that $N_x \neq N_y$. We may suppose that $N_x > N_{xy}$. Then $F = T \cap N(x)$ is a proper filter of T since $x \in F$, $y \notin F$, contrary to the hypothesis that T is $\mathfrak{N}$-simple.

The first statement of the following corollary justifies the terminology "$\mathfrak{N}$-simple."

II.2.17 COROLLARY. A semigroup S is $\mathfrak{N}$-simple if and only if S has a single $\mathfrak{N}$-class. The $\mathfrak{N}$-classes of S are precisely all maximal $\mathfrak{N}$-simple subsemigroups of S. The set N_x can be characterized either as the union of all $\mathfrak{N}$-simple subsemigroups of S containing x or as the greatest such subsemigroup.

Proof. Exercise.

II.2.18. Exercises

1. Show that trivial semigroups are the only semigroups S with the property that $N(x) = N_1(x)$ for all $x \in S$.

2. Establish the equivalence of the following conditions on a semigroup S.
   i) If $a \in SbScS$, then there exists a positive integer n for which $a^n \in ScbS$.
   ii) $N(x) = \{y \in S \mid x^n \in SyS$ for some positive integer $n\}$ for all $x \in S$.
   iii) For any $a,b,c,d \in S$ there exists a positive integer n for which $(abcd)^n \in SdcbaS$.

3. Let A be a subsemigroup of a semigroup S. Show that the set $\bigcup_{a \in A} N(a)$ is the least filter of S containing A.

4. Show that in a commutative semigroup S, an ideal I of S is completely prime if and only if for any ideals A,B of S, $AB \subseteq I$ implies that either $A \subseteq I$ or $B \subseteq I$.

5.* Let $S = \Pi_{i=1}^n S_i$. For $i = 1,2, \ldots ,n$, let F_i be a filter of S_i. Show that $F = \Pi_{i=1}^n F_i$ is a filter of S, and conversely, that every filter of S is of this form. Deduce that in S, $N_{(s_1,s_2, \ldots ,s_n)} = \Pi_{i=1}^n N_{s_i}$ and that $Y_S \cong \Pi_{i=1}^n Y_{S_i}$. Also show that the converse of the first statement is, in general, false for the direct product of an infinite number of semigroups.

6. Show that if T is a homomorphic image of S, then Y_T is a homomorphic image of Y_S.

7. For a nonempty set X, describe the greatest semilattice decomposition of the following semigroups.

i) The free semigroup on X.
ii) The free commutative semigroup on X.
iii) The free band on X.

8. A *groupoid* is a nonempty set together with a (not necessarily associative) multiplication. With the same definitions as for semigroups, prove the validity of 2.3 and 2.5 for groupoids. (Note that semilattice congruences on a groupoid have a different description from those on a semigroup.)

II.2.19 REFERENCES: Chrislock [2], Kist [1], Numakura [1], Petrich [1], [2], Šulka [1], Tamura [4], [5], [8], [9], [13], Tamura and Kimura [1], [2], Thierrin [1], [2], [4], Yamada [1].

II.3. Completely Semiprime Ideals and $\mathfrak{N}$-Subsets

The purpose of this section is two-fold: to characterize an arbitrary nonempty intersection of completely prime ideals, these being of particular interest in light of the preceding section, and to characterize subsets which are congruence classes of some semilattice congruence. To this end, the following concepts will prove convenient.

II.3.1 DEFINITION. Let S be a semigroup. An ideal I of S is *prime* if for any $a,b \in S$, $aSb \subseteq I$ implies that either $a \in I$ or $b \in I$; I is *semiprime* if for any $a \in S$, $aSa \subseteq I$ implies $a \in I$. A nonempty subset A of S is *completely semiprime* if for any $x \in S$, $x^2 \in A$ implies $x \in A$; A is an *m-system* if for any $a,b \in A$ there exists $x \in S$ such that $axb \in A$.

It is easy to see that every completely prime ideal is prime, that a proper ideal is prime if and only if its complement is an m-system, and that $x^n \in I$, where I is a completely semiprime ideal, implies $x \in I$. Further, a semiprime ideal is a generalization of a prime ideal just as a completely semiprime ideal is a generalization of a completely prime ideal. We will characterize completely semiprime ideals in several ways, particularly as intersections of completely prime ideals. For this we need the next four lemmas, in all of which S stands for an arbitrary semigroup. We start with semiprime ideals.

II.3.2 LEMMA. Let I be a semiprime ideal and M be an m-system such that $I \cap M = \varnothing$. Then there exists an m-system M^* maximal relative to the properties: $M \subseteq M^*$, $I \cap M^* = \varnothing$.

Proof. Let $\mathfrak{M}$ be the partially ordered set of all m-systems N such that $M \subseteq N$ and $I \cap N = \varnothing$. Then $\mathfrak{M} \neq \varnothing$ since $M \in \mathfrak{M}$. Let $\mathfrak{C}$ be a chain in $\mathfrak{M}$, and let $D = \bigcup_{C \in \mathfrak{C}} C$. If $a,b \in D$, then there exists $C \in \mathfrak{C}$ such that $a,b \in C$ so $axb \in C$ for some $x \in S$. Hence $axb \in D$ and thus D is an m-system. Clearly $M \subseteq D$ and $I \cap D = \varnothing$. Hence $D \in \mathfrak{M}$ and by Zorn's lemma, $\mathfrak{M}$ contains a maximal element M^*.

II.3.3 LEMMA. Let S, I, and M be as in 3.2, and let M^* be any m-system of S maximal relative to the properties: $M \subseteq M^*$, $I \cap M^* = \varnothing$. Then $S \backslash M^*$ is a minimal prime ideal of S containing I.

Proof. Let $\mathcal{P}$ be the partially ordered set of all ideals P of S such that $I \subseteq P$ and $M^* \cap P = \varnothing$. Similarly as above, Zorn's lemma implies that $\mathcal{P}$ contains a maximal element Q.

Let $a,b \notin Q$. Letting $A = J(a) \cup Q$, $B = J(b) \cup Q$, we see that A and B are ideals with the property $Q \subset A$, $Q \subset B$. Hence $I \subseteq A$, $I \subseteq B$, and by maximality of Q, we must have $A,B \notin \mathcal{P}$. Thus $M^* \cap A \neq \varnothing$, $M^* \cap B \neq \varnothing$; let $c \in M^* \cap A$, $d \in M^* \cap B$. Consequently, since M^* is an m-system, there exists $x \in S$ such that $cxd \in M^*$. Since $c \in M^* \cap (J(a) \cup Q)$ and $M^* \cap Q = \varnothing$, we must have $c \in J(a)$ so that $c = uav$ for some $u,v \in S^1$; similarly $d = zbw$ for some $z,w \in S^1$. If $aSb \subseteq Q$, then $cxd = ua(vxz)bw \in Q$ since Q is an ideal, and hence $cxd \in M^* \cap Q$, contradicting the fact that $M^* \cap Q = \varnothing$. By contrapositive we conclude that Q is a prime ideal and thus $Q' = S \backslash Q$ is an m-system if it is nonempty.

Since $Q \supseteq I$ and $Q \cap M^* = \varnothing$, it follows that $Q' \cap I = \varnothing$ and $M^* \subseteq Q'$. By maximality of M^*, we must have $Q' = M^*$ so that $S \backslash M^* = Q$. If J is a prime ideal satisfying $Q \supset J \supseteq I$, then letting $J' = S \backslash I$, we have that J' is an m-system such that $J' \cap \mathrm{I} = \varnothing$ and $J' \supset M^*$ contradicting maximality of M^*. Therefore $Q = S \backslash M^*$ is a minimal prime ideal of S containing I.

It is very easy to see that the nonempty intersection of prime ideals is a semiprime ideal; the next corollary gives the converse.

II.3.4 COROLLARY. Every semiprime ideal is the intersection of minimal prime ideals containing it.

Proof. Let I be a semiprime ideal of a semigroup S, and let $d \in S \backslash I$. Letting $M = \{d^n | n = 1,2, \ldots\}$, we obtain an m-system disjoint from I; by virtue of 3.2 and 3.3 there exists a minimal prime ideal P containing I and not containing d.

We are now ready to consider completely semiprime ideals.

II.3.5 LEMMA. Let I be a completely semiprime ideal of S. Then $a_1 a_2 \ldots a_n \in I$ implies $a_{1\pi} a_{2\pi} \ldots a_{n\pi} \in I$ for any permutation π of $\{1,2, \ldots, n\}$.

Proof. If $xy \in I$, then $(yx)^2 = y(xy)x \in I$ and thus $yx \in I$. We will use this and the hypothesis that I is a completely semiprime ideal repeatedly in the remainder of the proof.

Assume that $xyz \in I$. Then we see successively that each of the following elements belongs to I: yzx, $(yz)(xz)$, $(xz)(yz)$, $(yxzy)z$, $(zy)(xzy)$, $(xzy)^2$, xzy, yxz, zyx. Furthermore, the assumption implies directly that $zxy \in I$. Hence the assertion of the lemma is valid for $n = 2,3$.

Note that every permutation is a product of transpositions of adjacent factors. Hence in order to complete the proof, we need only show that $axyb \in I$ implies $ayxb \in I$. Assume that $axyb \in I$. Then we see successively that each of the following elements belongs to I: $(xyb)a, y(xyba)$, $(ba)(yxy)$, $(ba)(yx)^2$, $(yx)(ba)(yx)$, $(bayx)^2$, $b(ayx)$, $ayxb$.

II.3.6 LEMMA. Every prime ideal P minimal relative to containing a completely semiprime ideal I is completely prime.

Proof. We may suppose that $P \neq S$ so that $M = S \backslash P$ is an m-system. Let $a_0, a_1, a_2, \ldots, a_n \in M$. Then there exist elements $x_1, x_2, \ldots, x_n \in S$ such that

$$\begin{aligned} d_1 &= a_0 x_1 a_1 \in M, \\ d_2 &= d_1 x_2 a_2 \in M, \\ d_3 &= d_2 x_3 a_3 \in M, \\ &\cdots\cdots\cdots\cdots \\ d_n &= d_{n-1} x_n a_n \in M \end{aligned}$$

and thus

$$d_n = a_0x_1a_1x_2a_2 \ldots x_na_n \in M.$$

If $a_0a_1 \ldots a_n \in I$, then $d_n \in I$ by 3.5 and hence $M \cap I \neq \emptyset$, contradicting the hypothesis that $I \subseteq P = S\backslash M$. Hence the semigroup T generated by M is an m-system satisfying $M \subseteq T$, $T \cap I = \emptyset$. By 3.2 and 3.3, the minimality of P implies that M is an m-system maximal relative to the property of having empty intersection with I. But then $M = T$ which shows that M itself is a semigroup, and hence its complement P is a completely prime ideal.

We are now able to prove the desired result.

II.3.7 THEOREM. The following conditions on an ideal I of a semigroup S are equivalent.

i) I is the intersection of completely prime ideals.
ii) I is the intersection of minimal completely prime ideals containing it.
iii) I is the union of $\mathfrak{N}$-classes.
iv) I is completely semiprime.

Proof. i) *implies* ii). Let $I = \bigcap_{\alpha \in A} I_\alpha$ where each I_α is a completely prime ideal. Fix $\alpha \in A$ and let $\mathcal{I}$ be the partially ordered set of all completely prime ideals J of S for which $I \subseteq J \subseteq I_\alpha$. Then $\mathcal{I} \neq \emptyset$ since $I_\alpha \in \mathcal{I}$. Let $\mathfrak{C}$ be a chain in $\mathcal{I}$ and let $A = \bigcap_{C \in \mathfrak{C}} C$. Then $I \subseteq A \subseteq I_\alpha$ and A is an ideal of S. Since the partially ordered set $\{S\backslash C | C \in \mathfrak{C}\}$ forms a chain, it follows that $S\backslash A = \bigcup_{C \in \mathfrak{C}} (S\backslash C)$ is a subsemigroup of S or is empty, which shows that A is a completely prime ideal. Hence $A \in \mathcal{I}$ and the Minimal Principle assures the existence of a minimal element, say J_α, in $\mathcal{I}$. But then J_α is also minimal relative to the property of being a completely prime ideal containing I. In addition

$$I \subseteq \bigcap_{\alpha \in A} J_\alpha \subseteq \bigcap_{\alpha \in A} I_\alpha = I$$

so $I = \bigcap_{\alpha \in A} J_\alpha$.

ii) *implies* iii). This follows from 2.14.

iii) *implies* iv). If $x^2 \in I$, then $N_x = N_{x^2} \subseteq I$ so that $x \in I$.

iv) *implies* i). By 3.4, for every $d \notin I$, there exists a minimal prime ideal J containing I and not containing d. According to 3.6, J is completely prime. Thus I is the intersection of all completely prime ideals containing it.

II.3.8 COROLLARY. A semigroup is $\mathfrak{N}$-simple if and only if it contains no proper completely semiprime ideals.

This should be compared with 2.11, 2.13, and the first statement in 2.17. Note that in a semilattice every ideal is completely semiprime; compare the next corollary with 2.15.

II.3.9 COROLLARY. If I is a completely semiprime ideal of a semigroup S, then $J = \{N_x \in Y_S | \ x \in I\}$ is an ideal of Y_S. Conversely, if J is an ideal of Y_S, then $I = \{x \in S | \ N_x \in J\}$ is a completely semiprime ideal of S. This establishes a one-to-one inclusion preserving correspondence between the partially ordered set of all completely semiprime ideals of S and the partially ordered set of all ideals of Y_S.

Proof. Exercise.

We now turn our attention to the question: Which subsets of a semigroup can serve as a class of some semilattice congruence on S? For the sake of expediency, we introduce the following concept.

II.3.10 DEFINITION. A nonempty subset C of a semigroup S is an $\mathfrak{N}$-*subset* if C is completely semiprime and satisfies the condition: for any $x,y \in S$ and $z \in S^1$, $x,yz \in C$ implies $xy,zx \in C$.

II.3.11 LEMMA. Let C be an $\mathfrak{N}$-subset of a semigroup S. Then for any $a,b \in S$ and $x \in S^1$, $xab \in C$ implies $xba \in C$.

Proof. Suppose first that $c,d \in S$, $cd \in C$. Then $cd,cd \in C$ implies $d(cd),cd \in C$ whence $(dc)^2 = d(cd)c \in C$. Thus $cd \in C$ implies $dc \in C$; we will use this several times.

Next let $x,a,b \in S$ with $xab \in C$. Then $(ab)x,b(xa) \in C$ so that $a(bxb) = (ab)xb \in C$. Hence $(bxb)a,x(ab) \in C$ which implies $b(xbax) = (bxb)ax \in C$. Consequently $(xbax)b,a(bx) \in C$ so that $(xba)^2 = (xbax)ba \in C$ and hence $xba \in C$.

II.3.12 THEOREM. The following conditions on a nonempty subset C of a semigroup S are equivalent.

i) C is an $\mathfrak{N}$-subset.
ii) C is a class of a semilattice congruence.
iii) C is the intersection of a completely semiprime ideal and a filter.

Proof. i) *implies* ii). Define a relation σ on S by:

$$a \,\sigma\, b \quad \text{if for every} \quad x \in S^1, xa \in C \quad \text{if and only if} \quad xb \in C.$$

It is clear that σ is an equivalence relation and a left congruence. We will use 3.11 repeatedly and without express reference. Let $a \,\sigma\, b$ and $c \in S$. Let $x \in S^1$. If $x(ac) \in C$, then $(xc)a \in C$ so that $(xc)b \in C$ and thus $(xb)c \in C$. Consequently $x(ac) \in C$ implies $x(bc) \in C$ and conversely by symmetry. Hence σ is a right congruence and thus a congruence. Further, S/σ is commutative. Next let $x \in S$. If $xa \in C$, then $xa,ax \in C$ and hence $xa^2 \in C$ since C is an $\mathfrak{N}$-subset. Conversely, if $xa^2 \in C$, then $xa^2,xa^2 \in C$ which implies $xa^2x \in C$ and thus $(xa)^2 \in C$. But then $xa \in C$ since C is completely semiprime. Further, $a \in C$ if and only if $a^2 \in C$ since C is an $\mathfrak{N}$-subset. Consequently $a \,\sigma\, a^2$ which implies that S/σ is idempotent. Hence σ is a semilattice congruence. If $a \in C$ and $a \,\sigma\, b$, then $b \in C$, which implies that C is a union of σ-classes. Let $a,b \in C$. If $xa \in C$, then $b,xa \in C$ so that $bx \in C$ and thus also $xb \in C$; by symmetry, $xb \in C$ implies $xa \in C$. Hence $a \,\sigma\, b$, and we deduce that C is a σ-class.

ii) *implies* iii). Let C be a class of a semilattice congruence τ, let $S = \bigcup_{\alpha \in Y} S_\alpha$ be the corresponding decomposition of S, and let $C = S_\gamma$. Letting $A = \bigcup_{\alpha \leq \gamma} S_\alpha$ and $B = \bigcup_{\alpha \geq \gamma} S_\alpha$, it is easy to verify that A is a completely prime ideal and B is a filter such that $A \cap B = S_\gamma = C$.

iii) *implies* i). Let $C = A \cap B$ where A is a completely semiprime ideal and B is a filter. Since both A and B are completely semiprime, so is C. If $y,z \in S$ and $x,yz \in C$, then $xy,zx \in A$ and $x,y,z \in B$ so that $xy,zx \in B$ and thus $xy,zx \in C$. If $x,y \in C$, then clearly $xy,x \in C$. Hence C is an $\mathfrak{N}$-subset.

II.3.13 COROLLARY. Let x be an element of a semigroup S. Then N_x is the intersection of all $\mathfrak{N}$-subsets of S containing x.

Proof. Exercise.

II.3.14 COROLLARY. The following conditions on a subset C of a semigroup S are equivalent.

i) C is an $\mathfrak{N}$-class.
ii) C is a class of the greatest semilattice decomposition.
iii) C is a minimal $\mathfrak{N}$-subset.
iv) C is a maximal $\mathfrak{N}$-simple subsemigroup of S.
v) C is an $\mathfrak{N}$-simple $\mathfrak{N}$-subset.

Proof. Exercise.

II.3.15. Exercises

1. Let x be an element of a semigroup S. Let $P_1(x) = J(x)$; supposing that $P_n(x)$ has been defined for $n \geq 1$, let $P_{n+1}(x)$ be the ideal generated by all elements y of S for which $y^m \in P_n(x)$ for some m. Show that $P(x) = \bigcup_{n=1}^{\infty} P_n(x)$ is the least completely semiprime ideal of S containing x and is also the intersection of all completely prime ideals of S containing x, and that $N_x = N(x) \cap P(x)$. Deduce that if x is the zero of S, then $P(x)$ is the least completely semiprime ideal and an $\mathfrak{N}$-class of S.

2. Establish the equivalence of the following conditions on a semigroup S.
   i) Y_S is a chain.
   ii) The partially ordered set of completely prime ideals of S forms a chain.
   iii) Every completely semiprime ideal of S is completely prime.

3. Show that $\mathfrak{N}$-subsets of a semilattice Y coincide with convex subsemilattices of Y (a subset A of Y is *convex* if $a,b \in A$, $a < x < b$ implies $x \in A$).

4. Prove that the congruence σ constructed in the proof of 3.12 is the greatest congruence on S having C as one of its classes, and that either
   i) $Z = S/\sigma$ has a zero and the property that for any two distinct nonzero elements a and b of Z there exists $x \in Z$ such that exactly one of the products ax,bx is zero, or
   ii) $Z = S/\sigma$ is as in i) with a zero adjoined.

5. Show that in any semigroup S, an ideal I is prime if and only if for any ideals A, B of S, $AB \subseteq I$ implies that either $A \subseteq I$ or $B \subseteq I$.

6. For any semigroup S, let B_S and C_S be the greatest band and the greatest commutative homomorphic images of S, respectively. Show that the diagram

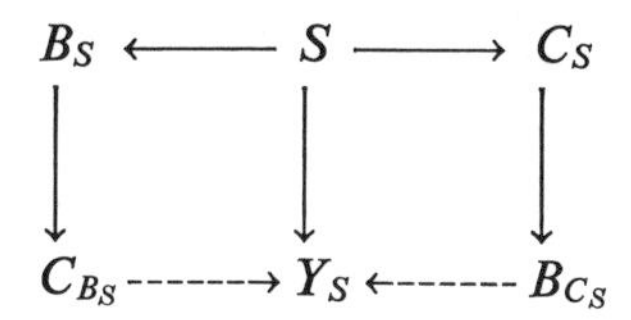

can be completed to a commutative diagram.

7.* Show that in any semigroup S, $SxS = Sx^2S$ for all $x \in S$ if and only if $N(x) = \{y \in S \mid SxS \subseteq SyS\}$ for all $x \in S$, and if so, every prime ideal of S is completely prime.

8. Show that an ideal of a semigroup is completely prime if and only if it is prime and completely semiprime.

9. Prove that every semigroup S with zero 0 such that 0 is a completely semiprime ideal of S is a subdirect product of semigroups with zero and without zero divisors.

II.3.16 REFERENCES: Adams [1], Andrunakievič and Rjabuhin [1], Grigor [1], Petrich [2], Szász [2], Thierrin [3].

II.4. Semilattices of Simple Semigroups

We are interested here in the conditions on a semigroup S equivalent to the requirement that every $\mathfrak{N}$-class of S be simple or satisfy a related but more restrictive condition. The pattern of the principal results in this section consists of establishing conditions equivalent to certain restrictions on all $\mathfrak{N}$-classes successively in terms of the following: ideals, elements, expressions for all $N(x)$, Green's relations, and ideals in conjunction with $\mathfrak{N}$-classes. In order to facilitate the discussion, we introduce some more notation and concepts. First recall I.3.6 and I.3.7:

$$L(x) = S^1x, \qquad R(x) = xS^1, \qquad J(x) = S^1xS^1$$

are the principal left, right, and two-sided ideals of a semigroup S generated by an element x of S.

II.4.1 DEFINITION. In any semigroup S, for $T = L,R,J$ define a relation $\mathfrak{J}$ by: $x \mathfrak{J} y$ if and only if $T(x) = T(y)$. The relations $\mathcal{L},\mathcal{R},\mathcal{H} = \mathcal{L} \cap \mathcal{R}$ and $\mathcal{J}$ are *Green's relations* (or *equivalences*) on S. For any Green's relation $\mathfrak{J}$, the $\mathfrak{J}$-class containing an element x of S is denoted by T_x (hence $T_x = \{y \in S | T(x) = T(y)\}$). A semigroup S is $\mathfrak{J}$-*simple* if it has only one $\mathfrak{J}$-class: for $\mathcal{L}$-simple we say *left simple*, similarly for *right simple*, and for $\mathcal{J}$-simple we say *simple* (see I.3.8).

Note that this terminology and notation conforms with the case $\mathfrak{J} = \mathfrak{N}$ introduced earlier. It is easy to see that, e.g., $x \mathcal{L} y$ if and only if for every left ideal L of S, either $x,y \in L$ or $x,y \notin L$. Further, in view of 3.7, $x \mathfrak{N} y$ if and only if for every completely semiprime ideal I of S, either $x,y \in I$ or $x,y \notin I$. Left, right, two-sided, and completely semiprime ideals are closed under union and nonempty intersection; filters only under nonempty intersection; completely prime ideals only under unions. Finally $\mathcal{H} \subseteq \mathcal{L} \subseteq \mathcal{J} \subseteq \mathfrak{N}$ where $\mathcal{L}$ can be substituted by $\mathcal{R}$. Also note that $\mathcal{L}$ is a right congruence and $\mathcal{R}$ is a left congruence.

In the following lemma and proposition, let P denote any property of a semigroup S which implies $\mathfrak{N}$-simplicity (e.g., simplicity, being a group, etc.). Recall the definition in 1.14.

II.4.2 LEMMA. If a semigroup S is a semilattice of semigroups Q_α each of which has property P, then the induced semilattice congruence equals $\mathfrak{N}$.

Proof. Since Q_α has property P, by 2.16 we have $Q_\alpha \subseteq N_x$ for some $x \in S$ which shows that the induced congruence is contained in $\mathfrak{N}$; the opposite inclusion follows by the minimality of $\mathfrak{N}$.

The next proposition will come in handy.

II.4.3 PROPOSITION. The following conditions on a semigroup S are equivalent.

i) Every $\mathfrak{N}$-class has property P.
ii) S is a semilattice of semigroups each of which has property P.

iii) For every $x \in S$, N_x is the greatest subsemigroup of S containing x and having property P.

Proof. That i) implies ii), and iii) implies i) is trivial; ii) implies i) by 4.2. Suppose that i) holds and let T be any subsemigroup of S containing x and having property P. Then T is $\mathfrak{N}$-simple and 2.16 implies that $T \subseteq N_y$ for some $y \in S$. But then obviously $T \subseteq N_y = N_x$ and thus iii) holds.

Hence in many results that follow, we may substitute a statement of type i) in 4.3 by a statement of type ii), while iii) gives an additional property of $\mathfrak{N}$-classes. The next lemma will be used without express mention.

II.4.4 LEMMA. Let x be an element of a semigroup S, then

$$N(x) = \bigcup_{N_y \geq N_x} N_y = \bigcup_{N_x = N_{xy}} N_y = \bigcup_{N_y \geq N_x} N(y).$$

Proof. Exercise.

We begin with simplicity of each $\mathfrak{N}$-class.

II.4.5 THEOREM. The following conditions on a semigroup S are equivalent.

i) Every $\mathfrak{N}$-class is simple.
ii) Every ideal of S is completely semiprime.
iii) For every $x \in S$, $x \in Sx^2S$.
iv) For every $x \in S$, $N(x) = \{y \in S \mid x \in SyS\}$.
v) $\mathfrak{N} = \mathcal{J}$.
vi) Every ideal of S is a union of $\mathfrak{N}$-classes.

Proof. i) *implies* ii). Let I be an ideal of S and let $x^2 \in I$. Then $x^2 \in I \cap N_x$; hence $I \cap N_x$ is an ideal of N_x and we must have $I \cap N_x = N_x$. But then $x \in I$ and I is completely semiprime.
ii) *implies* iii). For any $x \in S$, Sx^2S is an ideal of S and since $x^4 \in Sx^2S$, the hypothesis implies that $x \in Sx^2S$.
iii) *implies* iv). For $x \in S$, let

$$T = \{y \in S \mid x \in SyS\}.$$

The inclusion $T \subseteq N(x)$ is immediate since $N(x)$ is a filter. The hypothesis implies that $x \in T$. Thus to prove that $N(x) \subseteq T$, it suffices to show that T is a filter. Hence let $u,v \in T$. For any $a \in S$, we have

$$vau \in S(vau)^2S \subseteq SuvS$$

so that

$$x \in Sx^2S \subseteq S(SvS)(SuS)S \subseteq SuvS,$$

which shows that $uv \in T$. Conversely, if $uv \in T$, then it follows immediately that $u,v \in T$. Hence T is a filter and thus $T \subseteq N(x)$ and the equality prevails.

The implications "iv) *implies* v)" and "v) *implies* vi)" are trivial.

vi) *implies* i). Using I.3.9 it suffices to show that for any $x \in S$, $N_x \subseteq N_x y N_x$ for all $y \in N_x$. For $y,z \in N_x$, the hypothesis implies $N_x \subseteq J(y^3)$ since $y^3 \in N_x$, and thus $z = ay^3b$ for some $a,b \in S^1$. If $a \notin S$, then $N_{ay} = N_y = N_x$, and if $a \in S$, then

$$N_{ay} = N_a N_y = N_a N_z = N_a N_{ay^3b} = N_{ay^3b} = N_x.$$

Analogously $N_{yb} = N_x$ which implies

$$z = (ay)y(yb) \in N_x y N_x.$$

Consequently $N_x \subseteq N_x y N_x$ for all $y \in N_x$.

II.4.6 DEFINITION. A semigroup S satisfying the identity $x = xyx$ is a *rectangular band.*

It is easy to verify that a rectangular band is indeed a band (i.e., an idempotent semigroup).

II.4.7 COROLLARY. Every band S is a semilattice of rectangular bands, and for any $x \in S$,

$$N(x) = \{y \in S | \ x = xyx\}.$$

Proof. A band S trivially satisfies 4.5 iii), so by 4.5 iv) we have

$$N(x) = \{y \in S | \ x \in SyS\}.$$

Hence let $x = ayb$ for some $a,b \in S$. Then

$$\begin{aligned} x = aybx &= a(ybx)(ybx) = (aybx)(ybx) = x(ybx) \\ &= (xy)(xy)(bx) = xy(xybx) = xyx, \end{aligned}$$

which proves the last assertion of the corollary. The first assertion follows from the second, since the latter implies

$$N_x = \{y \in S \mid x = xyx,\ y = yxy\}$$

which is obviously a rectangular band.

II.4.8 COROLLARY. A band S is $\mathfrak{N}$-simple (simple) if and only if S is a rectangular band.

Proof. Exercise.

II.4.9 THEOREM. The following conditions on a semigroup S are equivalent.

i) Every $\mathfrak{N}$-class is left simple.
ii) Every left ideal of S is completely semiprime and two-sided.
iii) For every $x \in S$, $x \in Sx^2$ and $xS \subseteq Sx$.
iv) For every $x \in S$, $N(x) = \{y \in S \mid x \in Sy\}$.
v) $\mathfrak{N} = \mathfrak{L}$.
vi) Every left ideal of S is a union of $\mathfrak{N}$-classes.

Proof. i) *implies* ii). Let I be a left ideal of S. If $x^2 \in I$, then $x^2 \in I \cap N_x$; hence $I \cap N_x$ is a left ideal of N_x and we must have $I \cap N_x = N_x$. But then $x \in I$ and thus I is completely semiprime. If $x \in I$ and $y \in S$, then $yx \in I \cap N_{yx}$ so that $I \cap N_{yx} = N_{yx}$ since $I \cap N_{yx}$ is a left ideal of N_{yx}. But then $xy \in N_{yx}$ implies that $xy \in I$ and I is a two-sided ideal.
ii) *implies* iii). For any $x \in S$, Sx^2 is a left ideal of S and is thus completely semiprime. Since $x^3 \in Sx^2$, we have $x \in Sx^2$. Moreover, the set Sx is a left and thus a two-sided ideal of S and contains x, so that $xS \subseteq J(x) \subseteq Sx$.

The proof of the remaining implications is an easy modification of the corresponding proofs in 4.5 and is left as an exercise.

II.4.10 COROLLARY. The following conditions on a semigroup S are equivalent.

i) Every $\mathfrak{N}$-class is a group.
ii) Every left and every right ideal of S is completely semiprime and two-sided.
iii) For every $x \in S$, $x \in xSx$ and $xS = Sx$.
iv) For every $x \in S$, $N(x) = \{y \in S | x \in ySy\}$.
v) $\mathfrak{N} = \mathcal{H}$.
vi) Every left and every right ideal of S is a union of $\mathfrak{N}$-classes.

Proof. Exercise.

II.4.11 COROLLARY. The following conditions on a band S are equivalent.

i) Every $\mathfrak{N}$-class of S is a left zero semigroup.
ii) For every $x \in S$, $xS \subseteq Sx$.
iii) For every $x \in S$, $N(x) = \{y \in S | x = xy\}$.
iv) S satisfies the identity $xyx = xy$.

Proof. Exercise.

II.4.12 COROLLARY. The following conditions on a band S are equivalent.

i) S is a semilattice.
ii) For every $x \in S$, $xS = Sx$.
iii) For every $x \in S$, $N(x) = \{y \in S | x = yxy\}$.

Proof. Exercise.

II.4.13. Exercises

1. Prove that the following conditions on a semigroup S are equivalent.
   i) S is a chain of simple semigroups.

ii) Every ideal of S is completely prime.
iii) For any $x,y \in S$, either $x \in SxyS$ or $y \in SxyS$.

2.* Prove that the following conditions on a semigroup S are equivalent.
i) S is a chain of left simple semigroups.
ii) Every left ideal of S is two-sided and completely prime.
iii) For any $x,y \in S$, either $x \in Sxy$ or $y \in Syx$, and $xS \subseteq Sx$.

3. Find necessary and sufficient conditions in terms of elements of a semigroup S in order that S be a chain of rectangular bands. Do the same by substituting "left zero semigroup" instead of "rectangular band."

4. Prove that a semigroup S which is the union of simple subsemigroups is a semilattice of its maximal simple subsemigroups.

5. Let C be a commutative semigroup and T be a nontrivial rectangular band. Show that $C \times T$ is the only subdirect product of C and T contained in $C \times T$ if and only if C is a group.

6. Show that a semigroup S is a rectangular band if and only if for any $x,y \in S$, $xy = yx$ implies $x = y$.

7.* Prove that a semigroup S is a subdirect product of a commutative semigroup and a rectangular band if and only if
i) S satisfies the identity $axyb = ayxb$.
ii) For any $a,b,x,y \in S$, $xax = yby$ implies $xy = yx$.
iii) Denoting by $\mathcal{C}$ the least congruence σ on S for which S/σ is commutative, $x\mathcal{C}y$ and $xy = yx$ imply $x = y$.

II.4.14 REFERENCES: Croisot [1], Green [1], McLean [1], Petrich [2], Szász [2].

II.5. Weakly Commutative Semigroups

We have seen in the preceding section that for certain classes of semigroups, $N(x)$ has a particularly simple form. We will now consider yet another class of semigroups for which this holds. A condition somewhat weaker than commutativity is provided by the following.

II.5.1 DEFINITION. A semigroup S is *weakly commutative* if for any $x,y \in S$, $(xy)^n \in ySx$ for some positive integer n.

II.5.2 THEOREM. Let S be a semigroup. Then for every $x \in S$,

$$N(x) = \{y \in S | \, x^n \in ySy \text{ for some } n\}$$

if and only if S is weakly commutative.

Proof. Necessity. For any $x,y \in S$, we have $yx \in N(xy)$ which in view of the hypothesis yields $(xy)^n \in yxSyx \subseteq ySx$. Thus S is weakly commutative.

Sufficiency. For $x \in S$, let

$$T = \{y \in S \mid x^n \in Sy \text{ for some } n\}.$$

We show first that T is a filter of S. Let $y,z \in T$. Then $x^m = ay$ and $x^n = bz$ for some $a,b \in S$ and some m, n. Since S is weakly commutative, $(bz)^r = zc$ for some $c \in S$ and some r. Consequently

$$x^{m+nr} = (ay)(bz)^r = (ay)(zc)$$

whence, again by weak commutativity, we obtain $[(ayz)c]^k = d(ayz)$ for some $d \in S$ and some k. Hence

$$x^{(m+nr)k} = d(ayz) \in Syz,$$

and thus $yz \in T$. Conversely, suppose that $yz \in T$. Hence $x^m = (ay)z$ for some $a \in S$ and some m so that $z \in T$. By weak commutativity, we have $[(ay)z]^n = b(ay)$ for some $b \in S$ and some n, so

$$x^{mn} = (ba)y \in Sy$$

and hence $y \in T$. Therefore T is a filter, and since $x \in T$, by minimality of $N(x)$, we must have $N(x) \subseteq T$. On the other hand, $T \subseteq N_2(x) \subseteq N(x)$ so that $T = N(x)$.

By symmetry we can prove also that

$$N(x) = \{y \in S \mid x^m \in yS \text{ for some } m\},$$

which implies

$$\begin{aligned} N(x) &= \{y \in S \mid x^n \in Sy \text{ for some } n\} \cap \{y \in S \mid x^m \in yS \text{ for some } m\} \\ &= \{y \in S \mid x^k \in ySy \text{ for some k}\}. \end{aligned}$$

II.5.3 COROLLARY. If S is a weakly commutative semigroup, then for every $x \in S$,

$$N(x) = \{y \in S \mid x^n \in yS \text{ for some } n\}.$$

Proof. This was proved in the sufficiency part of 5.2.

II.5.4 COROLLARY. A weakly commutative band is a semilattice.

Proof. Exercise.

For (mainly commutative) semigroups it is convenient to have the following concept.

II.5.5 DEFINITION. A semigroup S is *archimedean* if for any $a,b \in S$ there exists a positive integer n for which $a^n \in SbS$.

In this terminology it is useful to keep the following consequence of 5.3 in mind.

II.5.6 COROLLARY. Every weakly commutative semigroup is (uniquely) a semilattice of archimedean semigroups.

II.5.7 DEFINITION. The archimedean subsemigroups in 5.6 of a weakly commutative semigroup S are the *archimedean components* of S (which obviously coincide with the $\mathcal{N}$-classes of S).

II.5.8. Exercises

1. In any semigroup S, define a relation $\mathcal{K}$ by:

   $x \,\mathcal{K}\, y$ if $x^m = y^n$ for some positive integers m, n $\quad (x,y \in S)$.

   Show that $\mathcal{K}$ is an equivalence relation contained in $\mathcal{N}$, and that each $\mathcal{K}$-class contains at most one idempotent. Also show that the congruence generated by $\mathcal{K}$ coincides with the least band congruence on S.

2. Denoting by K_x the $\mathcal{K}$-class of a semigroup S containing an element x, show that K_x contains an idempotent if and only if x is of finite order. If so, denoting the idempotent by e, show that

$$K_e = \{x \in S \mid x^n = e \text{ for some } n\}.$$

Deduce that every $\mathcal{K}$-class contains an idempotent (exactly one) if and only if S is periodic.

3.* Prove that $\mathcal{K} = \mathfrak{N}$ in a semigroup S implies that S is weakly commutative, and that the converse holds if S is periodic.

4.* Prove that in a periodic semigroup S the following conditions are equivalent.
   i) For any $e,f \in E_S$, $K_eK_f \subseteq K_{ef} = K_{fe}$.
   ii) For any $x,y \in S$, $(xy)^r = x^sy^t = y^tx^s$ for some positive integers r, s, t.
   iii) S is weakly commutative and E_S is commutative.
   iv) S is weakly commutative and E_S is a semigroup.

   Also show that under these conditions, for $x \in K_e$, $N(x) = \bigcup_{f \geq e} K_f$ and $Y_S \cong E_S$.

5. Show that the direct product of a weakly commutative archimedean semigroup and a rectangular band is $\mathfrak{N}$-simple.

6. Give an example of a weakly commutative semigroup which is neither commutative nor a group.

II.5.9 REFERENCES: Head [1], Petrich [2], Ponděliček [1], Schwarz [1], Sedlock [1], Tamura [15], Yamada [3].

II.6. Separative Semigroups

In this section we consider semigroups which are semilattices of cancellative semigroups. We then discuss the case of such semigroups which are also commutative. To begin, we introduce two new concepts.

II.6.1 DEFINITION. A semigroup S is *left cancellative* if for any $a,b,x \in S$, $xa = xb$ implies $a = b$; *right cancellative* if $ax = bx$ implies $a = b$; *cancellative* if it is both left and right cancellative.

A weakened version of a cancellative semigroup is provided by the following notion.

II.6.2 DEFINITION. A semigroup S is *separative* if for any $x,y \in S$,

i) $x^2 = xy$ and $y^2 = yx$ imply $x = y$,
ii) $x^2 = yx$ and $y^2 = xy$ imply $x = y$.

II.6.3 LEMMA. In a separative semigroup S, for any $x,y,a,b \in S$, the following statements hold.

i) $xa = xb$ if and only if $ax = bx$.
ii) $x^2a = x^2b$ implies $xa = xb$.
iii) $xya = xyb$ implies $yxa = yxb$.

Proof. i) If $xa = xb$, then $a(xa)x = a(xb)x$ and $b(xa)x = b(xb)x$ so that $(ax)^2 = (ax)(bx)$ and $(bx)^2 = (bx)(ax)$ which by separativity implies $ax = bx$. The opposite implication follows by symmetry.

ii) If $x^2a = x^2b$, then by part i), $xax = xbx$ whence $(ax)^2 = (ax)(bx)$ and $(bx)^2 = (bx)(ax)$ and thus by separativity $ax = bx$. But then by part i), also $xa = xb$.

iii) Let $xya = xyb$. Then $xyay = xyby$, and thus by part i), $yayx = ybyx$. Multiplying by suitable elements on the right and using part i), we obtain the following strings of equalities:

$$\begin{aligned} yayxa &= ybyxa & \qquad yayxb &= ybyxb \\ ayxay &= byxay & \qquad ayxby &= byxby \\ (ayx)^2 &= (byx)(ayx) & \qquad (ayx)(byx) &= (byx)^2 \end{aligned}$$

which by separativity implies $ayx = byx$. But then by part i) we have $yxa = yxb$.

II.6.4 THEOREM. A semigroup S is separative if and only if S is a semilattice of cancellative semigroups. If so, the relation σ defined on S by

$$x \,\sigma\, y \quad \text{if for any} \quad a,b \in S, \quad xa = xb \quad \text{if and only if} \quad ya = yb \tag{1}$$

is the greatest band congruence on S all of whose classes are cancellative.

Proof. *Necessity.* Let σ be as in (1). It is clear that σ is an equivalence relation. Let $x \,\sigma\, y$ and $z \in S$. It follows immediately that $xz \,\sigma\, yz$. If $(zx)a = (zx)b$, then $z(xa) = z(xb)$ which by

6.3 i) implies $(xa)z = (xb)z$. But then $x(az) = x(bz)$ so by (1), $y(az) = y(bz)$. Again by 6.3 i), it follows that $(zy)a = (zy)b$. By symmetry, $(zy)a = (zy)b$ implies $(zx)a = (zx)b$. Hence $zx \, \sigma \, zy$ and σ is a congruence. Further, 6.3 ii) implies that S/σ is a band, while 6.3 iii) implies that S/σ is commutative. Therefore σ is a semilattice congruence.

Suppose that $zx = zy$ and $x \, \sigma \, z$, $y \, \sigma \, z$. Since $x \, \sigma \, z$, $zx = zy$ implies $x^2 = xy$, and since $y \, \sigma \, z$, it implies $yx = y^2$. But then separativity yields $x = y$. If $xz = yz$ with $x \, \sigma \, z$, $y \, \sigma \, z$, then by 6.3 i), $zx = zy$, and this reduces to the case just considered. Hence each σ-class is cancellative.

Sufficiency. Let τ be a semilattice congruence on S all of whose classes are cancellative. Let $x,y \in S$. If $x^2 = xy$ and $y^2 = yx$, then $x \, \tau \, xy$ and $y \, \tau \, yx$ so that $x \, \tau \, y$, and thus the equation $x^2 = xy$ implies $x = y$. Similarly, $x^2 = yx$ and $y^2 = xy$ imply $x = y$. Therefore S is separative.

Let ξ be a band congruence on S all of whose classes are cancellative. Let $x \, \xi \, y$ and $xa = xb$. Then $x \, \xi \, y$ implies $xa \, \xi \, ya$ and $xb \, \xi \, yb$; the last equivalence yields $xa \, \xi \, yb$ since $xa = xb$. Consequently the three elements xa, ya, yb are all contained in the same ξ-class. Applying 6.3 iii) to $yxa = yxb$ yields $xya = xyb$ and again to $(ax)(ya) = (ax)(yb)$ yields $(xa)(ya) = (xa)(yb)$. Hence the cancellation law in the ξ-class containing xa, ya, yb implies $ya = yb$. By symmetry, we conclude that $ya = yb$ implies $xa = xb$ which shows that $x \, \sigma \, y$. Therefore $\xi \subseteq \sigma$.

II.6.5 COROLLARY. A semigroup S is separative if and only if every $\mathfrak{N}$-class of S is cancellative.

Proof. Exercise.

We will characterize the semigroups above in terms of subdirect products in III.7.6. Note that in the case of a commutative semigroup, separativity simplifies to: $x^2 = y^2 = xy$ implies $x = y$.

II.6.6 THEOREM. The following conditions on a semigroup S are equivalent.

i) S is commutative and separative.
ii) S is a semilattice of commutative cancellative semigroups.
iii) S is embeddable in a semilattice of abelian groups.

Proof. i) *implies* ii). This is a consequence of 6.4.

ii) *implies* iii). Let S be a semilattice of commutative cancellative semigroups S_α, $\alpha \in Y$. For $a \in S_\alpha$, $b \in S_\beta$, and $c \in S_{\alpha\beta}$, we obtain

$$\begin{aligned} c^2(ab) &= (ca)(bc) = (bc)(ca) \\ &= [(bc)c]a = c(bc)a = [(cb)c]a = c^2(ba) \end{aligned}$$

and thus $ab = ba$ by cancellation in $S_{\alpha\beta}$. Consequently S is commutative.

Let $F = \bigcup_{\alpha \in Y} (S_\alpha \times S_\alpha)$ and on F define a relation τ by

$$(a,b)\,\tau\,(c,d) \quad \text{if} \quad a,c \in S_\alpha \quad \text{for some} \quad \alpha \in Y \quad \text{and} \quad ad = bc.$$

It is straightforward to verify that τ is a congruence on F; let $Q = F/\tau$ and by $[a,b]$ denote the τ-class containing (a,b). Letting

$$G_\alpha = \{[a,b] \in Q \mid a \in S_\alpha\}$$

one easily sees that G_α is an abelian group, that Q is a semilattice of groups G_α, and that the mapping

$$\varphi{:}a \to [a^2,a] \qquad (a \in S)$$

is an embedding of S into Q. (Note that in the case that Y has only one element, this construction reduces to the familiar one of embedding a commutative cancellative semigroup into its group of fractions.)

iii) *implies* i). We may suppose that S is a subsemigroup of a semigroup T which is a semilattice of abelian groups G_α, $\alpha \in Y$. We have seen above that a semilattice of commutative cancellative semigroups is commutative, hence T and thus also S is commutative. Assume that $x^2 = y^2 = xy$ for $x,y \in S$. Then both x and y are contained in the same G_α which by cancellation yields $x = y$. Therefore S is also separative.

II.6.7 COROLLARY. A semigroup S is commutative and separative if and only if every $\mathfrak{N}$-class of S is commutative and cancellative (and thus also archimedean).

Proof. Exercise.

II.6.8. Exercises

1. Show that a cancellative semigroup S can have at most one idempotent and if so, then it is the identity of S.
2. Prove that a periodic cancellative semigroup is a group. Deduce that a periodic separative semigroup is a semilattice of groups.
3. Show that the free semigroup on a set is cancellative.
4. Show that any two idempotents of a separative semigroup commute.

II.6.9 REFERENCES: Burmistrovič [1], Gluskin [8], Hewitt and Zuckerman [1], McAlister [1], McAlister and O'Carroll [1].

II.7. $\mathfrak{N}$-Semigroups

We have seen in the preceding section that every commutative separative semigroup is a semilattice of commutative cancellative archimedean semigroups, and that the converse also holds. If a commutative cancellative archimedean semigroup S has an idempotent e, then by cancellation e is the identity of S, and since for any $a \in S$, there exists $x \in S$ such that $e = ax$, S is an abelian group. We will now elucidate the structure of the remaining commutative cancellative archimedean semigroups; it is convenient to give them a name.

II.7.1 DEFINITION. An *$\mathfrak{N}$-semigroup* is a commutative cancellative archimedean semigroup without idempotents.

Hence a commutative separative semigroup is a semilattice of $\mathfrak{N}$-semigroups and abelian groups and conversely.

II.7.2 LEMMA. For any elements a and b of a commutative archimedean semigroup S without idempotents we have $a \neq ab$.

Proof. Suppose that $a = ab$. The hypothesis implies that $b^n = ax$ for some positive integer n and some $x \in S$. Hence

$$a = ab = ab^2 = \dots = ab^n = a^2x$$

which implies that $(ax)^2 = ax$, contradicting the hypothesis that S has no idempotents.

The next theorem gives the structure of $\mathfrak{N}$-semigroups.

II.7.3 THEOREM. Let N be the additive semigroup of nonnegative integers, G be an abelian group, $I{:}G \times G \to N$ be a function satisfying:

i) $I(\alpha,\beta) + I(\alpha\beta,\gamma) = I(\alpha,\beta\gamma) + I(\beta,\gamma)$ $(\alpha,\beta,\gamma \in G)$,
ii) $I(\alpha,\beta) = I(\beta,\alpha)$ $(\alpha,\beta \in G)$,
iii) $I(\epsilon,\epsilon) = 1$ where ϵ is the identity of G,
iv) for each $\alpha \in G$ there exists $m > 0$ such that $I(\alpha^m,\alpha) > 0$.

On the set $S = N \times G$ define a multiplication by:

$$(m,\alpha)(n,\beta) = (m + n + I(\alpha,\beta),\alpha\beta).$$

Then S with this multiplication is an $\mathfrak{N}$-semigroup, to be denoted by (G,I). Conversely, every $\mathfrak{N}$-semigroup is isomorphic to some (G,I).

Proof. Let $S = (G,I)$ be as above.

1. S is associative. Using i) we obtain

$$\begin{aligned}[(m,\alpha)(n,\beta)](k,\gamma) &= (m + n + I(\alpha,\beta),\alpha\beta)(k,\gamma)\\ &= (m + n + I(\alpha,\beta) + k + I(\alpha\beta,\gamma),(\alpha\beta)\gamma)\\ &= (m + I(\alpha,\beta\gamma) + n + k + I(\beta,\gamma),\alpha(\beta\gamma))\\ &= (m,\alpha)(n + k + I(\beta,\gamma),\beta\gamma)\\ &= (m,\alpha)\,[(n,\beta)(k,\gamma)].\end{aligned}$$

2. S is commutative. This follows immediately from ii).
3. S is cancellative. Suppose that

$$(m,\alpha)(n,\beta) = (m,\alpha)(k,\gamma);$$

then $(m + n + I(\alpha,\beta),\alpha\beta) = (m + k + I(\alpha,\gamma),\alpha\gamma)$

so that $m + n + I(\alpha,\beta) = m + k + I(\alpha,\gamma), \qquad \alpha\beta = \alpha\gamma.$

Consequently, the last equation implies $\beta = \gamma$, and hence the first implies $n = k$. Thus $(n,\beta) = (k,\gamma)$.

4. S is archimedean. Given (m,α) and (n,β), we wish to show the existence of $p > 0$ and (k,γ) such that

$$(m,\alpha)^p = (n,\beta)(k,\gamma). \tag{1}$$

We will need the following identity:

$$I(\gamma,\epsilon) = 1 \qquad (\gamma \in G) \tag{2}$$

which follows from i) and iii) by setting $\alpha = \beta = \epsilon$. We consider two cases.

(a) $m > 0$. Then (2) yields $(m,\alpha) = (0,\epsilon)(m - 1,\alpha)$ so that to solve (1) it suffices to consider the case $(0,\epsilon)^p = (n,\beta)(k,\gamma)$. Choose p such that $p - 1 > n + I(\beta,\beta^{-1})$, let $k = p - 1 - n - I(\beta,\beta^{-1})$ and $\gamma = \beta^{-1}$. Then by (2),

$$(0,\epsilon)^p = (I(\epsilon^{p-1},\epsilon) + \ldots + I(\epsilon,\epsilon),\epsilon^p) = (p - 1,\epsilon),$$
$$(n,\beta)(k,\gamma) = (n,\beta)(p - 1 - n - I(\beta,\beta^{-1}),\beta^{-1}) = (p - 1,\epsilon),$$

so that $(0,\epsilon)^p = (n,\beta)(k,\gamma)$, as desired.

(b) $m = 0$. By iv), there exists $d > 0$ such that $I(\alpha^d,\alpha) > 0$. Hence $(m,\alpha)^{d+1} = (0,\alpha)^{d+1} = (s,\alpha^{d+1})$ where $s = I(\alpha^d,\alpha) + \ldots + I(\alpha,\alpha) > 0$. By case (a), $(m,\alpha)^{(d+1)p} = (n,\beta)(k,\gamma)$ for some $p > 0$ and $(k,\gamma) \in S$, as required.

Conversely, let S be an $\mathfrak{N}$-semigroup and fix $a \in S$. We will

5. S has no idempotents. This follows easily from iii).

use 7.2 without express reference.

1. Let $D = \bigcap_{n=1}^{\infty} a^n S$ and assume that $D \neq \varnothing$. Then D is a nonempty intersection of ideals and is hence an ideal. For $s \in S$ we have $a^m = sx$ for some $m > 0$ and some $x \in S$, and for any $d \in D$, we have $d = a^m y$ for some $y \in S$ so that

$$d = a^m y = (sx)y = s(xy) \in sS,$$

proving that $D \subseteq sS$ for all $s \in S$. It follows that D is contained in every ideal of S so in particular $D \subseteq dD$. Since the opposite inclusion is obvious, we have $D = dD$ for every $d \in D$. Thus for any $c,d \in D$, the equation $cx = d$ is solvable in D which together with commutativity implies that D is a group. But then D contains an idempotent contradicting the hypothesis that S has no idempotents. Therefore $D = \varnothing$.

2. For convenience we write $x = a^0 x$ for any $x \in S$. Now let

$$T_n = a^n S \backslash a^{n+1} S \qquad (n = 0,1,2,\ldots).$$

The fact that $D = \varnothing$ signifies that every element x of S can be written in the form $x = a^n z$ for some $z \notin aS$. In fact, both n and z are unique, since for $a^n z = a^m w$ with $z,w \notin aS$, $n \neq m$

would imply by cancellation that either $z \in aS$ or $w \in aS$. Thus $m = n$ so also $z = w$ by cancellation.

3. Let σ be a relation on S defined by:

$$x \,\sigma\, y \quad \text{if} \quad a^m x = a^n y \quad \text{for some} \quad m,n \geq 0.$$

It is easy to verify that σ is a congruence. We can thus denote the classes of σ by S_α where $\alpha \in$ G and G is a semigroup. It follows immediately that all positive powers of a constitute a single σ-class, to be denoted by S_ϵ. Further, $x \,\sigma\, a^n x$ for any $x \in S$ so that ϵ is the identity of G. For any $x \in S$ there exist $m > 0$ and $y \in S$ such that $a^m = xy$ which implies $xy \,\sigma\, a$ proving that the equation $\alpha\beta = \epsilon$ has a solution in G for any $\alpha \in G$. Consequently, G is a group.

4. We have seen in part 2 that every $x \in S$ can be uniquely written in the form $x = a^n z$ for some $z \in T_0 = S \backslash aS$. It follows that every x is σ-related to exactly one element of T_0 and thus T_0 intersects each σ-class S_α in a single element, say u_α. The set $\{u_\alpha | \, \alpha \in G\}$ now serves as a set of representatives of different σ-classes, and we have

$$S_\alpha = \{a^n u_\alpha \mid n \geq 0\} \qquad (\alpha \in G).$$

This suggests that every element of S is uniquely determined by a nonnegative integer n and an element α of G. To obtain the form of the product of two such elements in terms of this representation, it suffices to express the product of two representatives. For any $\alpha,\beta \in G$, we have $u_\alpha u_\beta \,\sigma\, u_{\alpha\beta}$ since σ is a congruence and thus $u_\alpha u_\beta = a^n u_{\alpha\beta}$ for some $n \geq 0$ since the other possibility, viz. $u_{\alpha\beta} = a^n u_\alpha u_\beta$, cannot occur for $n > 0$ because of $u_{\alpha\beta} \notin aS$. We thus define a function $I{:}G \times G \to N$ by the requirement

$$u_\alpha u_\beta = a^{I(\alpha,\beta)} u_{\alpha\beta} \qquad (\alpha,\beta \in G).$$

5. We now verify that $I(\alpha,\beta)$ has properties i) - iv).
To start with,

$$(u_\alpha u_\beta)u_\gamma = a^{I(\alpha,\beta)} u_{\alpha\beta} u_\gamma = a^{I(\alpha,\beta)+I(\alpha\beta,\gamma)} u_{\alpha\beta\gamma},$$
$$u_\alpha(u_\beta u_\gamma) = u_\alpha a^{I(\beta,\gamma)} u_{\beta\gamma} = a^{I(\beta,\gamma)+I(\alpha,\beta\gamma)} u_{\alpha\beta\gamma},$$

which by the uniqueness of the representation yields i). Item ii) is obvious. Since $u_\epsilon = a$, we immediately obtain iii). For any $\alpha \in G$, we have $u_\alpha^m \in aS$ since S is archimedean, and, on the other hand, $u_\alpha^m = a^{I(\alpha^{m-1},\alpha) + \ldots + I(\alpha,\alpha)} u_{\alpha^m}$ so $I(\alpha^k,\alpha) > 0$, establishing iv).

6. Define a mapping ψ by:

$$x\psi = (n,\alpha) \quad \text{if} \quad x = a^n u_\alpha \qquad (x \in S).$$

By the uniqueness of the representation, ψ is single-valued and one-to-one. For $x = a^m u_\alpha$ and $y = a^n u_\beta$, we obtain

$$\begin{aligned}(x\psi)(y\psi) &= (m,\alpha)(n,\beta) = (m + n + I(\alpha,\beta),\alpha\beta) \\ &= (a^{m+n+I(\alpha,\beta)} u_{\alpha\beta})\psi = (xy)\psi\end{aligned}$$

and ψ is a homomorphism. Since ψ obviously maps S onto (G,I), we conclude that $S \cong (G,I)$.

Note that $a\psi = (0,\epsilon)$. This shows that for every element $a \in S$, we can find an isomorphic copy of S of the form (G,I) for which $a\psi = (0,\epsilon)$. In the exercises below, we will see that for different elements a we get different groups G and thus also different functions I. We thus should write $(G,I)_a$ to express that this isomorphic copy of S depends on the choice of a. Hence this representation, and in particular the group G, is not an invariant of the semigroup S, which causes great difficulties in establishing when two (G,I) are isomorphic. Writing G_a for the group arising from a, it is natural to ask what all these groups have in common. Some of these properties will be considered in the next theorem, its corollary and the exercises, in particular the influence of such properties on S itself. Note that $(m,g) \to g$ is a homomorphism of (G,I) onto G. It is easy to verify that $(m,\alpha) = (0,\epsilon)^m(0,\alpha)$ for any $(m,\alpha) \in (G,I)$. Hence $\{(0,\alpha) \mid \alpha \in G\}$ is a set of generators of (G,I). In particular, if G is finite, then (G,I) is finitely generated. We now turn to the converse.

II.7.4 THEOREM. (G,I) is finitely generated if and only if G is finite.

Proof. Sufficiency was observed above. Assume that (G,I) is generated by a finite set T. Let T_0 be the subset of T consisting of elements of the form $(0,\alpha)$. Then $T_0 \neq \varnothing$, for otherwise no element of the form $(0,\alpha)$ could be expressed as a product of elements of T. Let $T_0 = \{(0,\alpha_i) \mid 1 \leq i \leq k\}$. For every $1 \leq i \leq k$, there exists a least positive integer m_i such that $I(\alpha_i^{m_i},\alpha_i) > 0$ (see 7.3). It follows that for $1 \leq s \leq m_i$, $(0,\alpha_i)^s$ is of the form $(0,\beta)$ for some $\beta \in G$, and for $s > m_i$, $(0,\alpha)^s$ is not of the form $(0,\beta)$ for any $\beta \in G$. Since T generates (G,I), for every element of the form $(0,\alpha)$, we have

$$(0,\alpha) = (0,\alpha_1)^{s_1}(0,\alpha_2)^{s_2} \dots (0,\alpha_k)^{s_k},$$

where $0 \leq s_i \leq m_i$, $i = 1,2, \dots, k$, and not all s_i are equal to zero ($s_i = 0$ means that $(0,\alpha_i)$ is absent). Consequently there exist only a finite number of elements of the form $(0,\alpha)$ which evidently implies that G is finite.

II.7.5 DEFINITION. The group G_a associated with an element a of an $\mathfrak{N}$-semigroup S is a *structure group* of S.

II.7.6 COROLLARY. The following conditions on an $\mathfrak{N}$-semigroup S are equivalent.

i) S is finitely generated.
ii) Some structure group of S is finite.
iii) Every structure group of S is finite.

Proof. Exercise.

The following concept represents a strengthening of the archimedean property.

II.7.7 DEFINITION. A semigroup S is *power joined* if for any elements a and b of S, there exist positive integers m and n for which $a^m = b^n$.

In the notation of 5.8, exercise 1, a semigroup is power joined if and only if it has a single $\mathcal{K}$-class. Further, a group is power joined if and only if it is periodic.

II.7.8. Exercises

1. Show that a finitely generated $\mathfrak{N}$-semigroup is power joined.
2.* Prove that the following conditions on (G,I) are equivalent.

i) G is periodic.
ii) (G,I) is power joined.
iii) Every group homomorphic image of (G,I) is periodic.
Deduce that if some structure group of (G,I) is periodic, then every structure group of (G,I) is periodic.

3. Let S be a commutative separative semigroup. Show that in S, $\mathcal{K} = \mathcal{N}$ (i.e., every archimedean component of S is power joined) if and only if every homomorphic image of S which is a semilattice of groups is periodic.

4.* Let S be a commutative power joined semigroup without idempotents. Fix $a \in S$ and define a function χ on S by:

$$\chi : x \longrightarrow \frac{m}{n} \quad \text{if} \quad x^n = a^m.$$

Show that χ is a homomorphism of S into the semigroup of positive rationals under addition, and that the congruence θ induced by χ is given by:

$$x \,\theta\, y \quad \text{if and only if} \quad x^n = y^n \quad \text{for some} \quad n,$$

and is thus independent of the choice of a.

5.* Find all (G,I) representations of the following semigroups under addition:
i) positive integers,
ii) positive rational numbers,
iii) positive real numbers.

6. Compute all functions I in 7.3 for G a group of order 2.

7. Let $S = (G,I)$ and let Z be the additive group of all integers. Using the function I, define a multiplication on the Cartesian product $H = Z \times G$ in such a way that H is a group isomorphic to the quotient group of S, with a subgroup K satisfying $K \cong Z$ and $H/K \cong G$.

II.7.9 REFERENCES: R. E. Hall [1], Higgins [1], [2], Levin [1], Levin and Tamura [1], McAlister [1], McAlister and O'Carroll [1], Petrich [3], Tamura [6], [12].

Ideal Extensions

If I is an ideal of a semigroup S, we may form the Rees quotient S/I which intuitively consists of elements of $S\backslash I$ and of a zero which replaces I. The semigroup S is said to be an ideal extension of I by S/I. We solve the ideal extension problem by starting with a semigroup I and a semigroup Q with zero and constructing all semigroups S such that I is an ideal of S and $Q \cong S/I$.

We approach this problem using left and right translations, and quickly note that this method is quite successful in case the future ideal is weakly reductive. We then construct explicitly several ideal extensions having special properties, such as strict, pure, retract, and dense extensions. A brief study is dedicated to an approach to this problem based on congruences rather than translations. Finally, for a semilattice Y and a semigroup S_α for each $\alpha \in Y$, we construct all semilattice compositions with these components.

III.1. Extensions and Translations

We introduce here some basic concepts related to ideal extensions. Those concerning the translational hull will be studied in greater detail in Chapter V.

III.1.1 DEFINITION. Let S be an ideal of a semigroup V. The relation τ defined on V by

$$x \tau y \quad \text{if either} \quad x,y \in S \quad \text{or} \quad x = y,$$

is a congruence called the *Rees congruence* induced by S. The quotient semigroup V/τ is the *Rees quotient semigroup* relative to S and is denoted by V/S.

We usually identify the congruence classes of τ different from S with the single element they contain. Thus we can informally construct the semigroup V/S by adjoining to the set $V \backslash S$ an element 0 not in $V \backslash S$ and on the set $(V \backslash S) \cup 0$ defining a multiplication $*$ by:

$$x * y = \begin{cases} xy & \text{if} \quad x,y,xy \notin S, \\ 0 & \text{otherwise}. \end{cases}$$

Since S is an ideal, the τ-class S, or equivalently the new element 0, acts as the zero of V/S.

III.1.2 DEFINITION. If S is an ideal of a semigroup V, then V is said to be an *ideal extension* of S by the Rees quotient semigroup V/S. Further V is a *proper extension* of S if $V \neq S$.

To simplify the terminology, we will often say that V is an ideal extension of S, or if no other kind of extension is considered, we will simply call V an *extension* of S.

To solve the extension problem for semigroups, we start with a semigroup S and a semigroup Q with zero (to avoid trivialities Q must have at least one nonzero element) and construct all semigroups V having an ideal S' such that $S \cong S'$, $V/S' \cong Q$. There is no harm in identifying S' with S and $Q^* = Q \backslash 0$, with $V \backslash S$, so we can take $V = S \cup Q^*$, of course assuming that S and Q^* are disjoint. The problem reduces to finding all associative multiplications on $V = S \cup Q^*$ which agree with existing products on S and Q^* and make S an ideal of V.

It is customary to denote the multiplication in V by a special symbol rather than by juxtaposition if S and Q are given and V is being constructed. However, if V is given, with multiplication denoted by juxtaposition as usual, or is being constructed but there is no danger of confusion with the multiplications in S or Q, we will have no special symbol for its multiplication.

Let S be an ideal of a semigroup V. Since S is an ideal of V, each $v \in V$ induces the following two functions on S:

$$\lambda^v : s \to vs, \qquad \rho^v : s \to sv \qquad (s \in S).$$

We write λ^v on the left and ρ^v on the right of its argument without parentheses and note that for all $x,y \in S$, $v,u \in V$,

$$\lambda^v(xy) = (\lambda^v x)y, \qquad (xy)\rho^v = x(y\rho^v)$$
$$x(\lambda^v y) = (x\rho^v)y, \qquad (\lambda^v x)\rho^u = \lambda^v(x\rho^u)$$

which is nothing but the associative law for certain triples in V. This observation motivates the introduction of the following concepts.

III.1.3 DEFINITION. Let S be a semigroup and let x and y denote arbitrary elements of S. A function λ on S is a *left translation* of S if $\lambda(xy) = (\lambda x)y$; a function ρ on S is a *right translation* of S if $(xy)\rho = x(y\rho)$. A left translation λ and a right translation ρ are *linked* if $x(\lambda y) = (x\rho)y$ in which case the pair (λ,ρ) is a *bitranslation* of S. We can consider a bitranslation (λ,ρ) as a "bioperator," i.e., as the function λ written on the left and as the function ρ written on the right, in which case it is convenient to use a single letter, say ω, and write ωx and $x\omega$ for any $x \in S$.

A left translation λ and a right translation ρ are *permutable* if $(\lambda x)\rho = \lambda(x\rho)$. A set T of bitranslations of S is *permutable* if for any $(\lambda,\rho),(\lambda',\rho') \in$ T, we have that λ and ρ' are permutable.

The set $\Lambda(S)$ of *all left translations* of S is a semigroup under the usual composition of functions: $(\lambda\lambda')x = \lambda(\lambda' x)$ for all $x \in S$; similarly the set $P(S)$ of *all right translations* of S is a semigroup under the composition: $x(\rho\rho') = (x\rho)\rho'$ for all $x \in S$. It is easy to verify that the product of two bitranslations is again a bitranslation, which leads us to the following fundamental concept.

III.1.4 DEFINITION. The subsemigroup of the direct product $\Lambda(S) \times P(S)$ consisting of all bitranslations of S is the *translational hull* of S, to be denoted by $\Omega(S)$.

Certain particularly interesting bitranslations are obtained as follows.

III.1.5 DEFINITION. Let s be an element of a semigroup S. Then the function λ_s defined by $\lambda_s x = sx$ for all $x \in S$ is the *inner left translation* induced by s; similarly the equation $x\rho_s = xs$ defines the *inner right translation* ρ_s induced by s; finally the pair $\pi_s = (\lambda_s, \rho_s)$ is the *inner bitranslation* induced by s. The set $\Pi(S)$ of all inner bitranslations is the *inner part of* $\Omega(S)$; we analogously define the sets $\Gamma(S)$ and $\Delta(S)$ of all inner left and of all inner right translations of S as the *inner part of* $\Lambda(S)$ and $\mathrm{P}(S)$, respectively.

III.1.6 LEMMA. In any semigroup S, we have

i) $\lambda\lambda_s = \lambda_{\lambda s}$, $\rho_s\rho = \rho_{s\rho}$ $\quad (s \in S, \lambda \in \Lambda(S), \rho \in \mathrm{P}(S))$,

ii) $\omega\pi_s = \pi_{\omega s}$, $\pi_s\omega = \pi_{s\omega}$ $\quad (s \in S, \omega \in \Omega(S))$.

Proof. With the notation as in the statement of the lemma and any $x \in S$, we obtain

$$(\lambda\lambda_s)x = \lambda(\lambda_s x) = \lambda(sx) = (\lambda s)x = \lambda_{\lambda s}x$$

proving that $\lambda\lambda_s = \lambda_{\lambda s}$, similarly for the second part of i). If $(\lambda,\rho) \in \Omega(S)$, then

$$x(\rho\rho_s) = (x\rho)\rho_s = (x\rho)s = x(\lambda s) = x\rho_{\lambda s}$$

proving that $\rho\rho_s = \rho_{\lambda s}$. This together with the first formula in i) establishes the first formula in ii); the second part of ii) is proved analogously.

III.1.7 COROLLARY. For any semigroup S, $\Gamma(S)$ is a left ideal of $\Lambda(S)$, $\Delta(S)$ is a right ideal of $\mathrm{P}(S)$, $\Pi(S)$ is an ideal of $\Omega(S)$.

Proof. Exercise.

It is easy to see that the mapping

$$\pi : s \to \pi_s \qquad (s \in S)$$

is a homomorphism. It is referred to as the *canonical homomorphism* from S into $\Omega(S)$ (or onto $\Pi(S)$).

Several kinds of restrictions on a semigroup occur often in connection with translations and extensions.

III.1.8 DEFINITION. A semigroup S is

i) *weakly reductive* if for any $a,b \in S$, $ax = bx$ and $xa = xb$ for all $x \in S$ imply $a = b$ (equivalently the canonical homomorphism $\pi: S \to \Omega(S)$ is one-to-one).

ii) *globally idempotent* if for every $a \in S$ there exist $x,y \in S$ such that $a = xy$ (usually written $S^2 = S$).

The next three lemmas will be quite useful.

III.1.9 LEMMA. Any two bitranslations of a weakly reductive semigroup S are permutable.

Proof. For any $x,y \in S$ and (λ,ρ), $(\lambda',\rho') \in \Omega(S)$, we obtain

$$x[(\lambda y)\rho'] = [x(\lambda y)]\rho' = [(x\rho)y]\rho' = (x\rho)(y\rho') = x[\lambda(y\rho')],$$
$$[(\lambda y)\rho']x = (\lambda y)(\lambda' x) = \lambda[y(\lambda' x)] = \lambda[(y\rho')x] = [\lambda(y\rho')]x$$

which by weak reductivity implies $(\lambda y)\rho' = \lambda(y\rho')$.

III.1.10 LEMMA. In a globally idempotent semigroup S, every left translation is permutable with every right translation.

Proof. For any $x,y \in S$ and $\lambda \in \Lambda(S)$, $\rho \in \mathrm{P}(S)$, we obtain

$$[\lambda(xy)]\rho = [(\lambda x)y]\rho = (\lambda x)(y\rho) = \lambda[x(y\rho)] = \lambda[(xy)\rho]$$

which by global idempotency implies $(\lambda s)\rho = \lambda(s\rho)$ for all $s \in S$.

III.1.11 LEMMA. If S is either a weakly reductive or a globally idempotent semigroup and $\omega,\omega' \in \Omega(S)$ are such that $\omega\pi_s = \omega'\pi_s$ and $\pi_s\omega = \pi_s\omega'$ for all $s \in S$, then $\omega = \omega'$.

Proof. With the given equations, we have $\pi_{\omega s} = \pi_{\omega' s}$ and $\pi_{\omega s} = \pi_{s\omega'}$ by 1.6, and in the case of weak reductivity, we obtain $\omega s = \omega' s$ and $s\omega = s\omega'$ for all $s \in S$. It further follows that for any $x,y \in S$,

$$\omega(xy) = (\omega x)y = (\omega' x)y = \omega'(xy)$$

and analogously $(xy)\omega = (xy)\omega'$, which in the case of global idempotency implies $\omega s = \omega' s$ and $s\omega = s\omega'$ for all $s \in S$.

We now return to extensions. The next result is simple but is of basic importance to this subject.

III.1.12 THEOREM. Let S be an ideal of a semigroup V. The mapping

$$\tau = \tau(V{:}S) : v \longrightarrow \tau^v = (\lambda^v, \rho^v) \qquad (v \in V),$$

where for each $v \in V$,

$$\lambda^v s = vs, \qquad s\rho^v = sv \qquad (s \in S),$$

is a homomorphism of V onto a semigroup of permutable bitranslations of S and extends the canonical homomorphism $\pi{:}S \longrightarrow \Omega(S)$. If S is weakly reductive or globally idempotent, then τ is the unique extension of π to a homomorphism of V into $\Omega(S)$.

Proof. That τ maps V onto a set of permutable bitranslations of S follows from the formulas preceding 1.3 which served as a motivation for the definitions involving translations. The homomorphism property of τ is easy to verify and is left as an exercise. From the very definitions, it is clear that τ extends π. Suppose next that $\omega{:}v \longrightarrow \omega^v$ is another homomorphism of V into $\Omega(S)$ extending π. Then for any $v \in V$ and $s \in S$, we obtain

$$\omega^v\pi_s = \omega^v\omega^s = \omega^{vs} = \pi_{vs} = \tau^{vs} = \tau^v\tau^s = \tau^v\pi_s$$

and dually $\pi_s\omega^v = \pi_s\tau^v$. Since this holds for all $s \in S$, we deduce by 1.11 that $\omega^v = \tau^v$, which then shows that $\omega = \tau$ whenever S is either weakly reductive or globally idempotent.

It is convenient to introduce two new concepts.

III.1.13 DEFINITION. The mapping $\tau(V{:}S)$ is the *canonical homomorphism* of V into $\Omega(S)$. The image of V under $\tau(V{:}S)$ is the *type* of the extension V of S and will be denoted by $\mathrm{T}(V{:}S)$.

Note that the canonical homomorphism $\pi{:}S \longrightarrow \Omega(S)$ is a special case of the one just introduced. If necessary, we will write $\tau^v(V{:}S)$ instead of τ^v.

III.1.14. Exercises

1. Show that if every element of a semigroup S has a left and a right identity, then every homomorphic image of S is weakly reductive, and that the converse holds if S is commutative.
2. Show that a finite commutative semigroup S is globally idempotent if and only if every homomorphic image of S is weakly reductive.
3. Show that a nontrivial semigroup S is globally idempotent if and only if no nontrivial homomorphic image of S is a zero semigroup.
4. Prove that a semigroup S is commutative, $\mathfrak{N}$-simple and contains an idempotent if and only if S is either an abelian group or an extension of an abelian group by a commutative nil semigroup. (A semigroup Q with zero is called *nil* if for every $q \in Q$ there exists a positive integer n such that $q^n = 0$.)
5. Show that a semigroup S is finite and cyclic if and only if S is either a finite cyclic group or an extension thereof by a nil cyclic semigroup.
6. Let S be a globally idempotent semigroup and an ideal of a semigroup T, and T be an ideal of a semigroup V. Show that S is an ideal of V.
7. Show that an extension V of a weakly reductive semigroup S by a weakly reductive semigroup Q is itself weakly reductive, but that the converse does not hold in general.
8. Prove that in a semigroup S every semilattice congruence is a Rees congruence induced by some ideal of S if and only if S is $\mathfrak{N}$-simple or an extension of an $\mathfrak{N}$-simple semigroup by a Kronecker semigroup.
9. Show that if S is a semigroup without idempotents, then $\Pi(S)$ has no idempotents.

III.1.15 REFERENCES: Clifford [2], Grillet and Petrich [1], [2].

III.2. Extensions of a Weakly Reductive Semigroup

We have seen in the preceding section that to find all extensions V of a semigroup S by a semigroup Q with zero and disjoint from S, we may take $V = S \cup Q^*$ where $Q^* = Q\backslash 0$, and on V find all associative multiplications which agree with the existing products in S and Q^* and make S an ideal of V. This actually gives an isomorphic copy of all semigroups V having S as an ideal satisfying $V/S \cong Q$.

The first result concerns extensions of an arbitrary semigroup. To state it we need two new concepts.

III.2.1 DEFINITION. Let S be a semigroup and Q be a semigroup with zero. A function $\theta : Q^* \to S$ is a *partial homomorphism* if $(ab)\theta = (a\theta)(b\theta)$ whenever $a,b,ab \in Q^*$. A function mapping the set $\{(a,b) \mid a,b \in Q^*, ab = 0\}$ into S is a *ramification function* of Q into S, to be denoted by $[\ ,\ ]:(a,b) \to [a,b]$.

III.2.2 THEOREM. Let S and Q be disjoint semigroups and let Q have a zero. Let $\theta : Q^* \to \Omega(S)$ be a partial homomorphism mapping Q^* onto a set of permutable bitranslations. and let $[\ ,\ ]$ be a ramification function of Q into S. We use the notation $\theta : a \to \theta^a = (\lambda^a, \rho^a)$ and assume:

(C1) $\qquad \theta^a\theta^b = \pi_{[a,b]} \qquad$ if $ab = 0$,

(C2) $\qquad [ab,c] = [a,bc] \qquad$ if $abc = 0,\ ab \neq 0,\ bc \neq 0$,

(C3) $\qquad [ab,c] = \lambda^a[b,c] \qquad$ if $ab \neq 0,\ bc = 0$,

(C4) $\qquad [a,bc] = [a,b]\rho^c \qquad$ if $ab = 0,\ bc \neq 0$,

(C5) $\qquad [a,b]\rho^c = \lambda^a[b,c] \qquad$ if $ab = bc = 0$.

On the set $V = S \cup Q^*$ define a multiplication $*$ by:

$$a * b = \begin{cases} a\rho^b & \text{if } a \in S,\ b \in Q^*, & \text{(M1)} \\ \lambda^a b & \text{if } a \in Q^*,\ b \in S, & \text{(M2)} \\ [a,b] & \text{if } a,b \in Q^*,\ ab = 0, & \text{(M3)} \\ ab & \text{otherwise.} & \text{(M4)} \end{cases}$$

Then V is an extension of S by Q, and conversely, every extension of S by Q can be so constructed.

Proof. The proof of the direct part consists of a straightforward verification of the associative law by considering various cases; the definition of multiplication clearly shows that S is an ideal of V. The converse is also easy to prove. For if V is an extension of

S, each element of V induces a bitranslation of S by the restriction of the inner bitranslation of V restricted to S, as we have seen in the preceding section. As for the ramification function, we define $[a,b] = ab$, the product in V, for any two elements $a,b \in Q^*$ for which $ab = 0$ in Q with $Q = V/S$. The conditions on the parameters θ and $[\ ,\ ]$ above as well as the formulas for multiplication follow without difficulty. A detailed proof is left as an exercise.

It takes only a little reflection to realize that the above theorem is nothing else but the disguised associative law in V and the requirement that S be an ideal of V. However, this theorem gives a general procedure for construction of extensions from which various special cases can be easily derived. It further shows that every extension of S by Q can be given by the two parameters θ and $[\ ,\ ]$, and we write

$$V = \langle S,Q;\theta,[\ ,\ ]\rangle.$$

In order to review all extensions of S by Q, one classifies them according to equivalence defined in some more or less arbitrary way. We will mention only one of these.

III.2.3 DEFINITION. For S a subsemigroup of semigroups V and V', a homomorphism of V into V' which leaves the elements of S fixed is an *S-homomorphism*; an *S-isomorphism* is an S-homomorphism which is also one-to-one and onto; an *S-endomorphism* of V is an S-homomorphism of V onto S. If S is an ideal of both V and V', and V and V' are S-isomorphic, then V and V' are *equivalent extensions* of S.

III.2.4 PROPOSITION. Two extensions $V = \langle S,Q;\theta,[\ ,\ ]\rangle$ and $V' = \langle S,Q';\theta',[\ ,\ ]'\rangle$ are equivalent if and only if there exists an isomorphism ψ of Q onto Q' such that $\theta = \psi\theta'$ and $[a,b] = [a\psi,b\psi]'$ for all $a,b \in Q^*$ for which $ab = 0$.

Proof. First let φ be an S-isomorphism of V onto V'. Define ψ on Q by: $\psi|_{Q^*} = \varphi|_{V \setminus S}$ and $0\psi = 0'$. If $a,b \in Q^*$ and $ab = 0$, then

$$[a,b] = ab = (ab)\varphi = (a\varphi)(b\varphi) = (a\psi)(b\psi) = [a\psi,b\psi]'.$$

Further, for any $a \in Q$ and $s \in S$, we obtain

$$\lambda^a s = as = (as)\varphi = (a\varphi)(s\varphi) = (a\psi)s = \lambda'^{a\psi} s$$

and analogously $s\rho^a = s\rho'^{a\psi}$, which together can be written as $\theta = \psi\theta'$. For the converse, we define φ to be the function on V which agrees with ψ on Q^* and is the identity map on S. It is left as an exercise to verify that φ is a homomorphism, all other properties being obvious.

Equivalent extensions should be considered as essentially the same. The two preceding results are simplified considerably in the following special cases: (i) Q has no zero divisors (equivalently, S is a completely prime ideal), in this case θ is a homomorphism and the ramification function and hence also conditions (C1)–(C5) are omitted; (ii) S is weakly reductive — for this case we have the following result.

III.2.5 THEOREM. Let S be a weakly reductive semigroup and Q be a semigroup with zero disjoint from S. Let $\theta: Q^* \to \Omega(S)$ be a partial homomorphism, in notation $\theta: a \to \theta^a = (\lambda^a, \rho^a)$, with the property that $\theta^a\theta^b \in \Pi(S)$ if $ab = 0$. Define a multipication $*$ on $V = S \cup Q^*$ by (M1), (M2), (M4) and

$$\text{(M3}') \qquad a * b = c \quad \text{where} \quad \theta^a\theta^b = \pi_c \quad \text{if} \quad a,b \in Q^*, \quad ab = 0.$$

Then V is an extension of S by Q, and conversely, every extension of S by Q can be so constructed.

Proof. Let θ be given as above. By 1.9 any two bitranslations of S are permutable. Because of the condition on θ and the weak reductivity of S, we may define a ramification function by the requirement

$$\pi_{[a,b]} = \theta^a\theta^b \quad \text{if} \quad a,b \in Q^*,\ ab = 0.$$

Then (C1) and (M3) in 2.2 are trivially satisfied. Let $a,b,c \in Q^*$. If $abc = 0$, $ab \neq 0$, $bc \neq 0$, we obtain

$$\pi_{[ab,c]} = \theta^{ab}\theta^c = \theta^a\theta^b\theta^c = \theta^a\theta^{bc} = \pi_{[a,bc]}$$

and (C2) holds. If $ab \neq 0$, $bc = 0$, using 1.6 we obtain

$$\pi_{[ab,c]} = \theta^{ab}\theta^c = \theta^a(\theta^b\theta^c) = \theta^a\pi_{[b,c]} = \pi_{\theta^a[b,c]} = \pi_{\lambda^a[b,c]}$$

and (C3) holds; (C4) is symmetric. Finally if $ab = bc = 0$, then using 1.6 twice we have

$$\pi_{[a,b]\rho^c} = \pi_{[a,b]\theta^c} = \pi_{[a,b]}\theta^c = \theta^a\theta^b\theta^c = \theta^a\pi_{[b,c]} = \pi_{\theta^a[b,c]} = \pi_{\lambda^a[b,c]}$$

proving (C5). By 2.2, V is an extension of S by Q.

Conversely, let V be an extension of S by Q. By 2.2, we have $V = \langle S,Q;\theta,[\ ,\ ]\rangle$ where θ trivially satisfies the conditions in the statement of the theorem.

The theorem and its proof show that in the case of a weakly reductive semigroup S, condition (C1) uniquely determines the ramification function, (C2)–(C5) are automatically satisfied, and in addition, any two bitranslations of S are permutable. This of course makes the theorem much more susceptible to successful applications than the theorem which deals with the general case. We will write in such a case $V = \langle S,Q;\theta\rangle$. It is convenient to introduce the following concept.

III.2.6 DEFINITION. Let S be a weakly reductive semigroup and Q be a semigroup with zero. A partial homomorphism $\theta: Q^* \to \Omega(S)$ such that $\theta^a\theta^b \in \Pi(S)$ if $a,b \in Q^*$, $ab = 0$ is an *extension function.*

We can sum up the theorem by saying that an extension of a weakly reductive semigroup S by a semigroup Q with zero is uniquely determined by an extension function $Q^* \to \Omega(S)$.

III.2.7 COROLLARY. Two extensions $V = \langle S,Q;\theta\rangle$ and $V' = \langle S,Q';\theta'\rangle$ of a weakly reductive semigroup S are equivalent if and only if there exists an isomorphism ψ of Q onto Q' such that $\theta = \psi\theta'$.

Proof. This follows easily from 2.5 and 2.4.

III.2.8 EXAMPLE. We will illustrate the theory above by constructing all extensions of the semigroup P of positive integers under addition by a semigroup Q with zero. For this we first need the translational hull of P. If $\lambda \in \Lambda(P)$, then for any $n > 1$, $\lambda n = \lambda(1 + (n - 1)) = \lambda 1 + n - 1$ which shows that λ is completely determined by its value at 1. Further, if $\lambda 1 = 1$, then λ is the identity mapping on P, otherwise $\lambda 1 = k > 1$ implies $\lambda = \lambda_{k-1}$. The same argument is valid for right translations. If $(\lambda,\rho) \in \Omega(S)$, then $(1\rho) + 1 = 1 + (\lambda 1)$ which shows that $\lambda 1 = 1\rho$. Hence $\Omega(S) \cong N$, the semigroup of nonnegative integers under addition. We identify $\Omega(S)$ with N, and immediately see that extension functions $\theta: Q^* \to N$ are given as follows: θ is a partial homomorphism such that $a\theta + b\theta > 0$ if $ab = 0$. By 2.5, the corresponding extension has the multiplication according to the formulas:

$$\begin{aligned} m * n &= n * m = m + n && \text{if} \quad m,n \in P, \\ n * a &= a * n = n + a\theta && \text{if} \quad n \in P,\, a \in Q^*, \\ a * b &= \begin{cases} a\theta + b\theta & \\ ab & \end{cases} && \begin{aligned} &\text{if} \quad a,b \in Q^*,\, ab = 0, \\ &\text{if} \quad a,b,ab \in Q^*. \end{aligned} \end{aligned}$$

By 2.7, two such extensions are equivalent if and only if there exists an isomorphism ψ of Q onto Q' such that $\theta = \psi\theta'$. In the case $Q \cong P^o$, θ is essentially an endomorphism of P. Such an endomorphism is uniquely determined by its value at 1, so we can write $\theta = \theta_k$ if $1\theta = k$, and hence $m\theta_k = km$. The mixed products are then given by:

$$m * a = a * m = m + k(a\varphi) \quad \text{if} \quad m \in P, \quad a \in Q^*,$$

where φ is an isomorphism of Q onto P^o. Since P has no automorphisms different from the identity mapping, φ is the unique isomporhism of Q onto P^o. Hence the extension is completely determined by the positive integer k. It is a simple exercise to show that for $k \neq k'$, the corresponding extensions are not equivalent. This gives a complete description of all extensions of an infinite cyclic semigroup by such a semigroup with a zero adjoined.

III.2.9. Exercises

1. Let V be an extension of a weakly reductive semigroup S by a semigroup Q with zero. Show that for any idempotents $e \in Q^*$ and $f \in S$, $e \geq f$ if and only i[illegible] $(V{:}S) \geq \pi_f$.

2. Show that if V is an extension of a commutative semigroup S, which is also either weakly reductive or globally idempotent, then S, is contained in the center of V.

3. Show that an extension of a weakly reductive commutative semigroup S by a commutative semigroup Q with zero is commutative.

4. Give an example of a noncommutative semigroup which is an extension of a commutative semigroup by a commutative semigroup.
5. Show that there exists no extension of an infinite cyclic semigroup by any Kronecker semigroup of order greater than 2.

III.2.10 REFERENCES: Clifford [2], McNeil [1], Petrich [8], Tamura [10], Verbeek [1], Yamada [8], Yamada and Tamura [1], Yoshida [1], Yoshida *et al.* [1].

III.3. Strict and Pure Extensions

For V an extension of S, we have defined in III.1 the type of the extension $T(V{:}S)$ as the image of V under the canonical homomorphism $\tau(V{:}S)$. We know that $T(V{:}S)$ always contains $\Pi(S)$ since $\tau(V{:}S)$ maps S onto $\Pi(S)$. This suggests the introduction of two kinds of extensions extremal relative to the properties of their types.

III.3.1 DEFINITION. An extension V of a semigroup S is *strict* if its type coincides with $\Pi(S)$; V is a *pure* extension of S if $\tau^a(V{:}S) \in \Pi(S)$ implies that $a \in S$.

It should first be noted that S is its only extension which is simultaneously strict and pure. We can rephrase these definitions as follows: V is a strict extension of S if and only if every element of V induces an inner bitranslation of S; V is a pure extension of S if and only if just the elements of S induce inner bitranslations of S. If we write $V = \langle S, Q; \theta, [\,,\,]\rangle$, then $\tau^a = \theta^a$ if $a \in V \backslash S$ and $\tau^a = \pi_a$ if $a \in S$. Thus V is a strict extension if and only if θ maps Q^* into $\Pi(S)$, pure extension if and only if θ maps Q^* into $\Omega(S) \backslash \Pi(S)$. The next result shows that strict and pure extensions occur naturally in every extension.

III.3.2 THEOREM. For an extension V of a semigroup S, the complete inverse image K of $\Pi(S)$ under $\tau(V{:}S)$ is the greatest strict extension of S contained in V; furthermore V is a pure extension of K.

Proof. Since $\tau(V{:}S)$ is a homomorphism and $T(V{:}S)$ contains $\Pi(S)$, K is an ideal of V, so V is indeed an extension of K. Since

$\tau(V:S)$ maps S onto $\Pi(S)$, we must have $S \subseteq K$, so K is an extension of S. It is obvious that K is then a strict extension of S. If V' is an extension of S and a subsemigroup of V, then $\tau(V':S)$ is the restriction of $\tau(V:S)$ to V'. Hence if V' is a strict extension of S, we must have $V' \subseteq K$. Let $v \in V$ and suppose that $\tau^v(V:K) \in \Pi(K)$. Then for some $a \in K$, we have $vk = ak$ and $kv = ka$ for all $k \in K$. This holds in particular for all $s \in S$, so

$$\tau^v(V:S) = \tau^a(V:S) = \tau^a(K:S) \in \Pi(S)$$

which implies that $v \in K$ proving that V is a pure extension of K.

III.3.3 COROLLARY. Every extension of S by Q is a pure extension of a strict extension. If Q has no proper nonzero ideals, an extension of S by Q is either strict or pure (recall from I.4 that Q is either 0-simple or a zero semigroup of order 2).

The above theorem can be used as an alternative way of constructing extensions if we are able to effectively construct these two special kinds of extensions. This seems to be possible in a simple enough form only for extensions of weakly reductive semigroups, and we now turn to this problem.

III.3.4 DEFINITION. Let S and S' be semigroups with zeros 0 and $0'$, respectively. A homomorphism φ of S into S' is *pure* if $s\varphi = 0'$ holds if and only if $s = 0$.

III.3.5 PROPOSITION. Let S be a weakly reductive semigroup and Q be a semigroup with zero. Let $\theta: Q \to \Omega(S)/\Pi(S)$ be a pure homomorphism. Then $V = \langle S, Q; \theta|_{Q^*}\rangle$ is a pure extension of S by Q, and conversely, every pure extension of S by Q can be so constructed.

Proof. In the direct part, we consider $\theta|_{Q^*}$ as a function from Q^* into $\Omega(S)\backslash\Pi(S)$; it then follows immediately from the hypothesis on θ that $\theta|_{Q^*}$ is an extension function which thus determines a pure extension of S. Conversely, let V be a pure extension of S by Q. Then by 2.5, we can write $V = \langle S, Q; \psi\rangle$ where $\psi: Q^* \to \Omega(S)$ is an extension function. Since the extension is pure, ψ maps Q^* into $\Omega(S)\backslash\Pi(S)$. We define $\theta: Q \to \Omega(S)/\Pi(S)$ to agree with ψ on Q^* and to map the zero of Q onto the zero of $\Omega(S)/\Pi(S)$. That θ is a pure homomorphism now follows from the hypothesis that the extension is pure and that ψ is an extension function. It is finally clear that $V = \langle S, Q; \theta|_{Q^*}\rangle$, as required. The details of this proof are left as an exercise.

The second part of our program, a construction of strict extensions of a weakly reductive semigroup, will be postponed until the next section where it will be considered in its natural frame of retract extensions.

III.3.6. Exercises

1. Let S be any semigroup and Q be a Kronecker semigroup such that $|Q| > |\Omega(S)/\Pi(S)|$. Show that there exists no pure extension of S by Q. Deduce also that if S has no idempotents, then there exists no extension of S by Q.
2. Let S be a semigroup without idempotents and Q be a semigroup which is either regular or periodic. Show that every extension of S by Q is pure.
3. Let S be a semigroup whose translational hull contains only one idempotent. For $n > 1$ and F a field, let Q be the set of all $n \times n$ matrices over F having at most one nonzero entry. Show that Q is a semigroup under multiplication of matrices and that every extension of S by Q is strict. Give an example of a semigroup satisfying the condition imposed on S.

III.3.7 REFERENCES: Grillet and Petrich [1].

III.4. Retract Extensions

The construction of strict extensions of a weakly reductive semigroup follows a pattern which yields a new kind of extension for an arbitrary semigroup. For this reason, we will first discuss this kind of extension for the general case, and then prove that, as a special case, we obtain all strict extensions of a weakly reductive semigroup. Recall 2.3, the definition of an S-endomorphism.

III.4.1 LEMMA. Let S be any semigroup, Q be a semigroup with zero disjoint from S, and $\eta: Q^* \to S$ be a partial homomorphism. On $V = S \cup Q^*$ define a multiplication $*$ by

$$a * b = \begin{cases} a(b\eta) & \text{if } a \in A,\ b \in Q^*, & \text{(M1}''\text{)} \\ (a\eta)b & \text{if } a \in Q^*,\ b \in S, & \text{(M2}''\text{)} \\ (a\eta)(b\eta) & \text{if } a,b \in Q^*,\ ab = 0, & \text{(M3}''\text{)} \\ ab & \text{otherwise.} & \text{(M4}'' = \text{M4)} \end{cases}$$

Then V is an extension of S by Q. Moreover, the function which is the identity on S and agrees with η on $V \backslash S$ is an S-endomorphism of V.

Proof. In order to use 2.2, we define θ and $[\ ,\ ]$ by:

$$\theta^a = \pi_{a\eta}, \qquad [a,b] = (a\eta)(b\eta) \quad \text{if} \quad ab = 0 \qquad (a,b \in Q^*).$$

Then θ maps Q^* into $\Pi(S)$, any two elements of which are permutable, and is a partial homomorphism since η is (indeed $\theta = \eta\pi$). For $a,b \in Q^*$ with $ab = 0$, we obtain

$$\theta^a\theta^b = \pi_{a\eta}\pi_{b\eta} = \pi_{(a\eta)(b\eta)} = \pi_{[a,b]}$$

and (C1) in 2.2 is satisfied. The verification of conditions (C2)–(C5) is just as simple and is left as an exercise. Multiplication formulas (M1)–(M4) evidently agree with (M1″)–(M4″) above. Hence V is an extension of S by Q by 2.2. The last assertion of the lemma follows immediately from the properties of η relative to the multiplication in V.

III.4.2 LEMMA. Let V be an extension of S by Q and let φ be an S-endomorphism of V. Then $\eta = \varphi|_{V \backslash S}$ is a partial homomorphism from Q^* into S, and the extension V is determined by η as in 4.1.

Proof. Exercise.

It is customary to call the extension V in 4.1 "determined by a partial homomorphism." However, in view of both 4.1 and 4.2, such extensions V coincide with those having an S-endomorphism. This suggests a somewhat less cumbersome terminology as follows.

III.4.3 DEFINITION. Let V be an extension of a semigroup S. Then S is a *retract ideal* of V, and V is a *retract extension* of S if V has an S-endomorphism; we may also say that its restriction to $V \backslash S = Q^*$ is a *partial homomorphism determining the extension.*

III.4.4 PROPOSITION. Every retract extension V of a semigroup S is strict. If S is weakly reductive, then the converse also holds and V has only one S-endomorphism.

Proof. The first statement follows from the first part of the proof of 4.1. If S is weakly reductive and V is a strict extension of S by Q, say $V = \langle S,Q;\theta \rangle$, then $\theta\pi^{-1}$ is a partial homomorphism of Q^* into S, and it is very easy to see that it determines the extension. The proof of the last statement of the proposition is left as an exercise.

Necessary and sufficient conditions for the validity of the converse will be established in 6.10. We now answer the following question: Which semigroups are retracts of every extension?

III.4.5 PROPOSITION. A semigroup S is a retract of every extension if and only if S has an identity.

Proof. Necessity. In particular, S must be a retract of the extension V which consists of S and an identity e adjoined to it. If φ is an S-endomorphism of V, then for every $s \in S$, we have

$$s(e\varphi) = s * e = s = e * s = (e\varphi)s$$

proving that $e\varphi$ is the identity of S.

Sufficiency. Let V be an extension of S. Let 1 be the identity of S, and define φ by: $v\varphi = v * 1$ for all $v \in V$. Then φ leaves S elementwise fixed and for any $x,y \in V$, we have

$$\begin{aligned}(x * y)\varphi &= (x * y) * 1 = x * (y * 1) = x * [1 * (y * 1)] \\ &= (x * 1) * (y * 1) = x\varphi * y\varphi,\end{aligned}$$

so φ is an S-endomorphism of V, and V is a retract extension of S.

Since retract extensions can, in general, be more easily constructed than many other kinds of extensions, it is of interest to know whether a given extension is a retract extension by the criteria involving special elements or subsets of the extension. We will prove only one such result for the important class of regular semigroups.

III.4.6 LEMMA. Every regular semigroup is weakly reductive.

Proof. Exercise.

The desired theorem can now be established.

III.4.7 THEOREM. Let V be an extension of a regular semigroup S by a semigroup Q with zero each of whose elements has either an idempotent left or right identity. Then V is a retract extension of S if and only if every element of Q^* has an idempotent left or right identity e such that the set $M_e = \{f \in E_S \mid f < e\}$ admits **a** greatest element.

Proof. Necessity. Let φ be an S-endomorphism of V and let $e \in E_{Q^*}$. Then

$$e * (e\varphi) = (e\varphi)(e\varphi) = (e\varphi) * e = e\varphi$$

and thus $e\varphi \in M_e$. If $f \in M_e$, then

$$f(e\varphi) = f * e = f = e * f = (e\varphi)f,$$

which implies that $f \leq e\varphi$. Thus $e\varphi$ is the greatest element of M_e.

Sufficiency. We now denote multiplication in V by juxtaposition. We show first that $M_e \neq \varnothing$ for every idempotent e in Q^*. Let e be such an idempotent and let $P = eSe$. Then P is a subsemigroup of S. If $x \in P$, then $x = eye$ for some $y \in S$ and $x = xzx$ for some $z \in S$ since S is regular. Consequently

$$x = (eye)z(eye) = (eye)(eze)(eye) = x(eze)x,$$

where $eze \in P$. This says that P is regular, and hence P contains an idempotent, say f. But then $f = ewe$ for some $w \in S$, so $f = ef = fe$, i.e., $f < e$, and therefore $f \in M_e$. Denote the greatest element of M_e by $\bar{e}$.

Next let $a \in Q^*$, and let e be an idempotent left identity of a (the case in which we assume that e is a right identity is proved similarly), and let x be an arbitrary element of S. Then $ax = (ax)u(ax)$ for some $u \in S$ by the regularity of S. Let $t = axue$; then

$$t^2 = (axue)(axue) = (axu)(axu)e = (axu)e = t$$

and clearly $t < e$ so that $t \leq \bar{e}$. Hence

$$\begin{aligned} ax &= (ax)u(ax) = (axue)ax = t(ax) \\ &= (\bar{e}t)(ax) = \bar{e}(tax) = \bar{e}(ax). \end{aligned} \tag{1}$$

Moreover, $xe = (xe)v(xe)$ for some $v \in S$. Let $s = evxe$; then

$$s^2 = (evxe)(evxe) = e(vxe)(vxe) = e(vxe) = s,$$

and clearly $s < e$ so that $s \leq \bar{e}$. Hence

$$\begin{aligned} xa &= (xe)a = (xevxe)a = x(evxe)a = xsa = x(s\bar{e})a \\ &= x(evxe)\bar{e}a = (xe)\bar{e}a = x(e\bar{e})a = x(\bar{e}a). \end{aligned} \tag{2}$$

From (1) and (2), respectively, we obtain

$$\lambda^a x = ax = \bar{e}ax = \lambda_{\bar{e}a}x \qquad (x \in S),$$

$$x\rho^a = xa = x(\bar{e}a) = x\rho_{\bar{e}a} \qquad (x \in S).$$

Hence $\tau^a = \pi_{\bar{e}a} \in \Pi(S)$. Consequently the extension V is strict and hence a retract extension by 4.6 and 4.4.

The restriction on Q in the above theorem is a mild one; it is satisfied in a regular semigroup (for if $a = axa$, then ax and xa are idempotent left and right identities of a, respectively), or in a semigroup having a left or right identity e, in which case, it suffices to require that M_e have a greatest element for this element alone. The next construction provides a simple way of extending a semigroup to a "very similar" larger semigroup.

III.4.8 DEFINITION. To every element s of a semigroup S associate a set Z_s such that: the sets Z_s are pairwise disjoint and $Z_s \cap S = \{s\}$. The set $V = \bigcup_{s \in S} Z_s$ together with the multiplication $x * y = ab$ if $x \in Z_a$, $y \in Z_b$, is an *inflation* of S. The function $\xi: V \to S$ which for all $s \in S$, maps Z_s onto s is an *inflation function*.

III.4.9 PROPOSITION. A semigroup V is an inflation of a semigroup S if and only if V is a retract extension of S by a zero semigroup Q.

Proof. For the direct part, it is easily verified that every inflation function is an S-endomorphism, and the form of the multiplication obviously implies that Q is a zero semigroup. Conversely, if ξ is an S-endomorphism and Q is a zero semigroup, then letting $Z_s = \{v \in V \mid v\xi = s\}$, one verifies without difficulty that ξ is an inflation function. The details of this proof are left as an exercise.

III.4.10. Exercises

1. Let V be an extension of a weakly reductive globally idempotent semigroup S by a semigroup Q with zero. Construct the translational hull of V in terms of the translational hulls of S and Q and find its inner part.

2. Prove that every extension of a semigroup S is strict if and only if S has an identity.

3.* Prove that a semigroup S for which S^2 is a semilattice of groups is an inflation of S^2. (*Hint:* Show that $a^2 \in G_e$, $b^2 \in G_f$ implies $ab \in G_{ef}$.)

4. Show that a semigroup S is an inflation of a band (semilattice, rectangular band, left zero semigroup, respectively) if and only if it satisfies the identity $xy = x^2y^2 = (xy)^2$ ($xy = yx = x^2y$, $xyz = xz$, $xy = x^2$, respectively).

5.* Let S be a regular semigroup, Q be a 0-simple semigroup. Let V be an extension of S by Q and assume that there exists an idempotent e in Q^* for which the set $M_e = \{f \in E_S \mid f < e\}$ has a greatest element. Prove that V is a retract extension of S by Q.

6. Let φ be an I-endomorphism of a semigroup S for some ideal I of S. Show that φ written on the left (right) is a left (right) translation of S and that the two are linked and permutable.

7. Let S be a regular semigroup. Show that the set of all I-endomorphisms of S, where I runs over all retract ideals of S, is a semilattice with identity under the composition of transformations.

III.4.11 REFERENCES: Arendt and Stuth [2], Petrich [5], [7], [10], Putcha and Weissglass [1], Schwarz [2], Tamura [14], Tully [1].

III.5. Dense Extensions

There is yet another kind of extension which appears quite frequently. While the definitions of strict and pure extensions are based on the translational hull and that of a retract extension on the notion of a partial homomorphism, the concept of a dense extension is based upon congruences.

III.5.1 DEFINITION. Let S be an ideal of a semigroup V. A congruence σ on V is an *S-congruence* if its restriction to S is the equality relation on S. Let $\mathfrak{Z} = \mathfrak{Z}(V{:}S)$ be the congruence on V induced by $\tau = \tau(V{:}S)$. If σ is any congruence on V, the set $\sigma(S)$ which is the union of all σ-classes having a nonempty intersection with S is the *saturation* of S by σ. If $\sigma(S) = S$, we say that S is *saturated for* σ.

III.5.2 THEOREM. Let S be an ideal of a semigroup V. Then every S-congruence on V is contained in $\mathfrak{Z} = \mathfrak{Z}(V{:}S)$, and $\mathfrak{Z}(S)$ is the largest strict extension of S contained in V. If S is weakly reductive, then $\mathfrak{Z}$ is the greatest S-congruence on V.

Proof. Let σ be an S-congruence on V. Then for any $a,b \in V$ and $s \in S$, the relation $a \,\sigma\, b$ implies $as \,\sigma\, bs$ and $sa \,\sigma\, sb$. But then $as = bs$ and $sa = sb$ since σ is an S-congruence, and thus $\tau^a(V{:}S) = \tau^b(V{:}S)$ proving that $\sigma \subseteq \mathfrak{Z}(V{:}S)$. The second assertion follows immediately from the first statement in 3.2. The third statement follows from the first and the fact that $\tau(V{:}S)|_S$ is one-to-one if S is weakly reductive.

III.5.3 COROLLARY. An extension V of a weakly reductive semigroup S is pure if and only if S is saturated for every S-congruence on V.

Proof. Exercise.

We now come to the basic concept of this section.

III.5.4 DEFINITION. An extension V of a semigroup S is *dense* if the equality relation is the only S-congruence on V.

III.5.5 COROLLARY. An extension V of a weakly reductive semigroup S is dense if and only if $\tau(V:S)$ is one-to-one.

Proof. If V is a dense extension of S, then $\mathfrak{I}(V:S)$ must be the equality since by 5.2 it is an S-congruence, so $\tau(V:S)$ is one-to-one. Conversely, if $\tau(V:S)$ is one-to-one, then $\mathfrak{I}(V:S)$ is the equality relation. But by 5.2 we know that $\mathfrak{I}(V:S)$ contains all S-congruences, so the extension must be dense.

Note that for any S and V, $\tau(V:S)$ is one-to-one if and only if no two different elements of V induce the same bitranslation on S. In such a case, S is automatically weakly reductive, and we see that, by identifying $\Pi(S)$ with S, dense extensions are, up to an isomorphism, simply all subsemigroups of $\Omega(S)$ containing $\Pi(S)$.

III.5.6 COROLLARY. Let S be a weakly reductive subsemigroup of a semigroup V. Then V is a dense extension of S if and only if there exists an isomorphism φ of V into $\Omega(S)$ extending the canonical homomorphism $\pi:S \rightarrow \Omega(S)$.

Proof. The direct part follows from 5.5 by taking $\varphi = \tau(V:S)$. Conversely, if φ is such an isomorphism, then S is an ideal of V since $\Pi(S)$ is an ideal of $\Omega(S)$, and $\varphi = \tau(V:S)$ by 1.12 so the extension is dense by 5.5.

III.5.7 COROLLARY. If V is a dense extension of a weakly reductive semigroup S, then every extension of S contained in V is also dense.

Proof. For any extension V' of S contained in V, we have that $\tau(V':S)$ is the restriction of $\tau(V:S)$ to V' and hence must be one-to-one by 5.5 if V is a dense extension.

We are now ready for another fundamental concept.

III.5.8 DEFINITION. An ideal S of a semigroup V is a *densely embedded ideal* if V is under inclusion a maximal dense extension of S.

For the present time, we will limit ourselves only to the following result dealing with this notion.

III.5.9 THEOREM. A weakly reductive semigroup S is a densely embedded ideal of a semigroup V if and only if $\tau(V:S)$ is an isomorphism of V onto $\Omega(S)$.

Proof. Let S be a densely embedded ideal of V. Hence V is a dense extension of S and thus $\tau = \tau(V:S)$ is one-to-one by 5.5. If τ were not onto, then we could in a natural way define a multiplication on $V \cup (\Omega(S)\backslash T(V:S))$ making it a dense extension of S strictly containing V, contradicting the maximality of V.

Conversely, let τ be an isomorphism of V onto $\Omega(S)$. By 5.5, V is a dense extension of S. Let V' be any extension of S strictly containing V as a subsemigroup, and let $a \in V'\backslash V$. Then $\tau^a(V':S) \in \Omega(S)$ and since $\tau(V:S)$ maps V onto $\Omega(S)$, there exists $b \in V$ such that $\tau^b(V':S) = \tau^b(V:S) = \tau^a(V':S)$. Since $a \neq b$, it follows from 5.5 that V' is not a dense extension of S. Consequently V is a maximal dense extension of S.

In light of 1.12, we can also say that an extension V of a weakly reductive semigroup S is a (maximal) dense extension if and only if there exists an isomorphism of V into (onto) $\Omega(S)$ extending the canonical homomorphism $\pi:S \to \Omega(S)$.

III.5.10 LEMMA. If V and V' are extensions of a semigroup S, and φ is an S-homomorphism of V into V', then $\tau(V:S) = \varphi[\tau(V':S)]$.

Proof. Exercise.

III.5.11 PROPOSITION. Two equivalent extensions of any semigroup have the same type. Conversely, any two dense extensions of a weakly reductive semigroup having the same type are equivalent.

Proof. The first statement follows immediately from 5.10. For the converse, assume that V and V' are dense extensions of a weakly reductive semigroup S and that $T(V:S) = T(V':S)$. Then $\varphi = \tau(V:S)\,[\tau(V':S)]^{-1}$ is clearly an S-isomorphism of V onto V' and these two extensions are equivalent.

III.5.12 COROLLARY. The classes of equivalent extensions of a weakly reductive semigroup S can be put in a one-to-one correspondence with the types of S, and consequently, with the set of all subsemigroups of $\Omega(S)/\Pi(S)$ containing its zero. In particular, if S is a densely embedded ideal of the semigroups V and V', then V and V' are S-isomorphic.

We will now cast a brief look at the following question: Which properties of a semigroup S carry over to all its dense extensions? Several weak "cancellative type" properties, somewhat stronger than weak reductivity, will come in handy.

III.5.13 DEFINITION. A semigroup S is *left reductive* if for any $a,b \in S$, $sa = sb$ for all $s \in S$ implies $a = b$; *right reductive* if $as = bs$ for all $s \in S$ implies $a = b$; *reductive* if it is both left and right reductive.

III.5.14 PROPOSITION. A dense extension of a left reductive right cancellative semigroup is left reductive and right cancellative.

Proof. Let V be a dense extension of a left reductive right cancellative semigroup S. Suppose that $va = vb$ for all $v \in V$. Then $sa = sb$ for all $s \in S$ and hence $s(at) = s(bt)$ for all $s,t \in S$. Since $s,at,bt \in S$, left reductivity of S implies $at = bt$. Consequently $\tau^a = \tau^b$ which by density implies $a = b$. Hence V is left reductive.

Suppose next that $ac = bc$. Then $(sa)(ct) = (sb)(ct)$ for all $s,t \in S$. Since $sa,ct,sb \in S$, right cancellativity in S yields $sa = sb$. We have seen above that $sa = sb$ for all $s \in S$ implies $a = b$. Hence V is right cancellative.

III.5.15 COROLLARY. Let V be an extension of a semigroup S. Then V is cancellative if and only if S is cancellative and the extension is dense.

Proof. Necessity. If $\tau^a = \tau^b$, then for any $s \in S$, $sa = sb$ so that $a = b$; S must obviously be cancellative.

Sufficiency. This is a consequence of 5.14.

III.5.16 PROPOSITION. A dense extension of a commutative reductive semigroup is commutative and reductive.

Proof. Let V be a dense extension of a commutative reductive semigroup S. Let $a,b \in V$ and suppose that $va = vb$ for all $v \in V$. In particular, $sa = sb$ for all $s \in S$. For any $s,t \in S$, we obtain

$$s(at) = (sa)t = (sb)t = s(bt)$$

so $at = bt$ for all $t \in S$. It follows that $\tau^a = \tau^b$ and thus $a = b$. Now let $x,y \in S$, $a,b \in V$. Then

$$\begin{aligned} x(yab) &= [x(ya)]b = [(ya)x]b = (ya)(xb) \\ &= (xb)(ya) = [(xb)y]a = [y(xb)]a \\ &= (yx)ba \;\;= (xy)ba \;\;= x(yba) \end{aligned}$$

which by reductivity yields $yab = yba$. We have seen above that then $ab = ba$.

III.5.17 COROLLARY. A dense extension of a semilattice is a semilattice.

Proof. Let V be a dense extension of a semilattice S. Then 5.16 implies that V is commutative. For any $v \in V$, $s \in S$, we have

$$sv = (sv)^2 = s^2v^2 = sv^2$$

so $\tau^v = \tau^{v^2}$ and thus $v = v^2$.

III.5.18 COROLLARY. Let V be an extension of a semigroup S. Then V is commutative and cancellative if and only if S is commutative and cancellative and the extension is dense.

Proof. Exercise.

III.5.19. Exercises

1. Let S be any semigroup and V be the semigroup obtained by the adjunction of an identity to S. Show that V is either a retract or a

dense extension of S. Deduce that if S has no identity, then S^1 is a dense extension of S.

2. Show that if V is an extension of S such that $\tau(V:S)$ is one-to-one, then the extension is dense. Deduce that for a globally idempotent semigroup S, $\Omega(S)$ is a dense extension of $\Pi(S)$.

3. Let D be a dense extension of a semigroup S, and V be both an extension of S and a dense extension of D. Show that V is a dense extension of S.

4.* Let S be a semigroup, $(\lambda,\rho) \in \Omega(S)\backslash\Pi(S)$, P be the semigroup generated by $\Pi(S) \cup (\lambda,\rho)$, and let $Q = P/\Pi(S)$. Prove that there exists an extension V of S by Q for which $(\lambda,\rho) * x = \lambda x$, $x * (\lambda,\rho) = x\rho$ for all $x \in S$ if and only if λ and ρ are permutable and if $(\lambda,\rho)^n \in \Pi(S)$ for some n, then for the smallest such there exists an element a of S such that $(\lambda,\rho)^n = \pi_a$, $\lambda a = a\rho$. Also show that such an extension is dense, and conversely, if S is weakly reductive, then up to equivalence, every dense extension of S by a cyclic semigroup with a zero adjoined can be so constructed.

5. Prove the following statements concerning an extension V of a nontrivial semigroup S.
   i) If V is subdirectly irreducible, then the extension is dense.
   ii) If S is subdirectly irreducible and the extension is dense, then V is subdirectly irreducible.
   iii) If V is subdirectly irreducible and every congruence on S is the restriction of some congruence on V, then S is subdirectly irreducible.

6. Prove that an extension V of a weakly reductive semigroup S is pure if and only if there exists an S-homomorphism φ of V onto a dense extension D of S such that $S\varphi^{-1} = S$.

7.* Let S be a semigroup and $Q = G^0$, a group G with a zero adjoined. Prove that there exists an extension of S by Q which is a cancellative semigroup if and only if S is a cancellative semigroup without idempotents and G is isomorphic with a subgroup of the group of units of $\Omega(S)$.

8. Let S be a weakly reductive semigroup without identity whose translational hull contains only one idempotent, and let Q be a semigroup with zero containing at least two nonzero idempotents whose product is zero. Show that there exists no extension of S by Q. Give an example of a class of semigroups satisfying the restrictions placed upon S and Q, respectively.

III.5.20 REFERENCES: Gluskin [1], [2], [3], [4], [5], Grillet and Petrich [1], Heuer [1], Heuer and Miller [1], Kalmanovič [1], Ljapin [1], [3], Petrich [16], Ševrin [1], [3].

III.6. Extensions of an Arbitrary Semigroup

We have seen in 2.2 how to construct any extension of an arbitrary semigroup S. However, the conditions in that theorem clearly indicate that the construction in question consists of a convenient way of writing the associative law in the extension and the requirement that S be an ideal. We have also seen that this approach is quite fruitful when S is weakly reductive, as exhibited by 2.5 and the succeeding results. For the general case, in order to get away from mere paraphrasing the associative law, we will now devise a different approach to extensions based on congruences.

III.6.1 LEMMA. Let V be an extension of a semigroup S. Let σ be an S-congruence on V and consider V/σ as an extension of S. Then V/σ is a dense extension of S if and only if σ is a maximal S-congruence on V.

Proof. The proof is an easy application of I.5.14 and is left as an exercise.

The next theorem gives a general means for constructing an extension of an arbitrary semigroup by using a dense extension of such a semigroup.

III.6.2 THEOREM. Let D be an extension of a semigroup S, and Q be a semigroup with zero disjoint from S. Further let $\varphi: Q^* \to D$ be a partial homomorphism for which $(a\varphi)(b\varphi) \in S$ whenever $ab = 0$ in Q. Let $V = S \cup Q^*$ with a multiplication defined by

$$a * b = \begin{cases} (a\varphi)b & \text{if} \quad a \in Q^*, b \in S, \\ a(b\varphi) & \text{if} \quad a \in S, b \in Q^*, \\ (a\varphi)(b\varphi) & \text{if} \quad a,b \in Q^*, ab = 0 \text{ in } Q, \\ ab & \text{otherwise.} \end{cases}$$

Then V is an extension of S, to be denoted by $[S,Q;\varphi,D]$. Conversely, every extension of S by Q can be constructed in this fashion from some extension D of S, which, in addition, may be assumed to be dense and satisfy the condition $D = S \cup Q^*\varphi$.

Proof. The proof of the direct part consists of a simple verification of the associative law by considering different cases and is left as an exercise. The converse is trivial if we do not impose any restrictions on D, for in such a case we can always take $D = V$ and φ the identity mapping. However, we can construct a D with the above properties as follows. Let V be an extension of S by Q. In the partially ordered set of S-congruences on V, a standard Zorn's lemma argument shows that there exists a maximal S-congruence σ on V. Let $D = V/\sigma$, let $\nu:V \to V/\sigma$ be the natural homomorphism, and let φ be the restriction of ν to $V \backslash S = Q^*$. If $a,b \in Q^*$ and $ab \neq 0$ in Q, then $(a\varphi)(b\varphi) = (a\nu)(b\nu) = (ab)\nu = (ab)\varphi$ and φ is a partial homomorphism. If $a,b \in Q^*$ and $ab = 0$ in Q, then $ab \in S$ in V and thus $(a\varphi)(b\varphi) = (a\nu)(b\nu) = (ab)\nu = ab \in S$, as required. Further, we have $D = V\nu = S \cup Q^*\varphi$, and D is a dense extension of S by 6.1. If $a \in S$, $b \in Q$, then in V, $ab = (ab)\nu = (a\nu)(b\nu) = a(b\varphi)$; the remaining cases are verified just as easily. This proves that V can be obtained from D as in the converse part of the theorem and that D satisfies all the additional requirements.

Note that for a weakly reductive semigroup S, identifying $\Pi(S)$ with S, we can take $D = \Omega(S)$; 6.2 then reduces to 2.5. This theorem reduces the construction of all extensions of an arbitrary semigroup S to a determination of all dense extensions of S. Unfortunately, except in the weakly reductive case, the latter are not available. Nevertheless, the above result somewhat clears up the situation concerning extensions of an arbitrary semigroup.

III.6.3 LEMMA. Let σ be an S-congruence on an extension V of S. Letting $\nu:V \to V/\sigma$ be the natural homomorphism, we have

$$\tau(V{:}S) = \nu \cdot \tau(V/\sigma{:}S), \qquad \mathrm{T}(V{:}S) = \mathrm{T}(V/\sigma{:}S).$$

Proof. For any $v \in V$ and $s \in S$, we obtain

$$vs = (vs)\nu = (v\nu)(s\nu) = (v\nu)s$$

and analogously $sv = s(v\nu)$ proving the first formula. The second formula follows from the first.

III.6.4 COROLLARY. Every type of extension is a type of a dense extension.

Proof. Exercise.

III.6.5 PROPOSITION. Every extension V of a semigroup S by a semigroup Q with zero is a subdirect product of D and Q, where D is a dense extension of S of the same type as V.

Proof. Let σ be a maximal S-congruence on V. Then $D = V/\sigma$ is a dense extension of S by 6.1 and has the same type as V by 6.3. Let $\nu:V \rightarrow V/\sigma$ and $\mu:V \rightarrow V/S$ be the natural homomorphisms, and define $\psi:v \rightarrow (v\nu,v\mu)$ $(v \in V)$. Then $\psi:V \rightarrow D \times Q$ is a homomorphism. If for $a,b \in V$, we have $a\psi = b\psi$, then $a\nu = b\nu$ and $a\mu = b\mu$ and we distinguish the following cases: (i) if $a,b \in S$, then $a\nu = b\nu$ implies that $a = b$, (ii) if $a,b \in V\backslash S$, then $a\mu = b\mu$ implies that $a = b$, (iii) if, e.g., $a \in S$ and $b \in V\backslash S$, then $a\mu = 0$ and $b\mu \neq 0$ which is impossible. Thus ψ is one-to-one, and since both ν and μ are onto, the image $V\psi$ is a subdirect product of D and Q.

III.6.6 COROLLARY. Every retract extension V of a semigroup S by a semigroup Q with zero is a subdirect product of S and Q.

Proof. If φ is an S-endomorphism of V, then the induced congruence is an S-congruence, maximal by 6.1 since S is its own dense extension. We may now apply the proof of 6.5.

III.6.7 COROLLARY. Every extension V of a weakly reductive semigroup S by a semigroup Q with zero is a subdirect product of the type of V and Q. Hence $\Omega(S) \times Q$ contains, up to equivalence, all extensions of S by Q.

Proof. The first statement follows immediately from 5.5 and 6.5; the second follows from the first.

Weakly reductive semigroups S can also be characterized by the property of a greatest S-congruence in every extension as follows.

III.6.8 PROPOSITION. A semigroup S is weakly reductive if and only if in every extension V of S there exists a greatest S-congruence.

Proof. If S is weakly reductive and V is an extension of S, then by virtue of 5.2, $\mathfrak{Z}(V:S)$ is the greatest S-congruence on V. Conversely, suppose that S is not weakly reductive. Then for some $a,b \in S$, we have $\pi_a = \pi_b$, $a \neq b$. Let $Q = \{0,c\}$ be a zero semigroup, and let V be the extension determined by the partial homomorphism $\varphi:c \to a$. Since $\pi_a = \pi_b$, V is also determined by the partial homomorphism $\psi:c \to b$. Extending φ and ψ to all of V by letting $s\varphi = s\psi = s$ for all $s \in S$, we obtain two S-homomorphisms of V. Hence the induced congruences are S-congruences which are clearly maximal. Thus V has no greatest S-congruence.

The next result elucidates the relationship among various kinds of extensions. For this we need a new concept.

III.6.9 DEFINITION. Using the notation of 4.8, let T be a subset of S. Then V is an *inflation of S over T* if for all $a \in S\backslash T$, $Z_a = \{a\}$. (Roughly speaking, only the part T of S may be inflated.)

III.6.10 THEOREM. The following conditions on a semigroup S are equivalent.

i) No proper dense extension of S is strict.
ii) Every strict extension of S is a retract extension.
iii) S is an inflation of a weakly reductive semigroup R over $R\backslash R^2$.
iv) For any $x,y \in S$, $\pi_x = \pi_y$ and $x \in S^2$ imply $x = y$.

Proof. i) *implies* ii). Let V be a strict extension of S and let σ be a maximal S-congruence on V. Then by 6.1, V/σ is a dense extension of S; by 6.3, $\mathrm{T}(V/\sigma:S) = \mathrm{T}(V:S)$ so that V/σ is a strict extension of S. The hypothesis then implies that $V/\sigma = S$. Consequently, every element v of V is σ-related to a unique element $\bar{v}$ of S, and the mapping $v \to \bar{v}$ is an S-homomorphism, proving that V is a retract extension of S.

ii) *implies* iii). Let Q be the semigroup $\Pi(S)$ with a zero adjoined, and consider the extension $V = \langle S,Q;\theta\rangle$ where θ is the identity mapping on $\Pi(S)$. This extension has type $\Pi(S)$ so it is strict and thus, by hypothesis, is determined by a partial homomorphism, say φ. Since Q has no zero divisors, φ is a homomorphism and by the definition of this extension, we have

$$(\pi_x\varphi)y = \pi_x * y = \lambda_x y = xy,$$

and dually, $y(\pi_x\varphi) = yx$ for all $x,y \in S$. This implies that $\pi_{\pi_x\varphi} = \pi_x$ for all $x \in S$. Since π is an onto mapping, $\varphi\pi$ is the identity mapping on $\Pi(S)$. Hence φ is one-to-one and $\alpha = \pi\varphi$ is an idempotent homomorphism of S into S. The image R of S under α is also the image of $\Pi(S)$ under φ and $R \cong \Pi(S)$ since φ is one-to-one.

To prove that S is an inflation over R, first observe that $\pi_{x\alpha} = \pi_{\pi_x\varphi} = \pi_x$ for all $x \in S$. Consequently for all $x,y \in S$,

$$xy = \lambda_x y = \lambda_{x\alpha} y = (x\alpha)y = (x\alpha)\rho_y = (x\alpha)\rho_{y\alpha} = (x\alpha)(y\alpha), \quad (1)$$

$$(x\alpha)y = [(x\alpha)y]\alpha = (x\alpha^2)(y\alpha) = (x\alpha)(y\alpha) = xy. \quad (2)$$

We will now prove that R is weakly reductive by showing that $\Pi(S)$ has this property (recall that $R \cong \Pi(S)$). Let $\pi_y,\pi_z \in \Pi(S)$ be such that

$$\pi_x\pi_y = \pi_x\pi_z, \qquad \pi_y\pi_x = \pi_z\pi_x \qquad (x \in S).$$

Since $\alpha = \pi\varphi$, we obtain

$$\begin{aligned} xy &= (x\alpha)(y\alpha) = (\pi_x\varphi)(\pi_y\varphi) = \pi_{xy}\varphi = \pi_{xz}\varphi \\ &= (\pi_x\varphi)(\pi_z\varphi) = (x\alpha)(z\alpha) = xz, \end{aligned}$$

and dually, $yx = zx$ for all $x \in S$. Thus $\pi_y = \pi_z$.

We now show that S is an inflation over $R\backslash R^2$. Take $a \in S$ such that $a\alpha \in R^2$. The proof will be completed when we show that $a \in R$. By hypothesis $a\alpha = bc$ for some $b,c \in R$. Let b',c' be two elements not contained in S. Let $V = S \cup b' \cup c'$ with the multiplication $*$ defined by

$$x * y = xy,$$

$$x * b' = xb, \qquad b' * x = bx, \qquad x * c' = xc, \qquad c' * x = cx,$$

$$b' * b' = b^2, \qquad c' * c' = c^2, \qquad c' * b' = cb, \qquad b' * c' = a,$$

for all $x,y \in S$. To prove the associativity of $*$, let $x,y,z \in V$. The equation $(x * y) * z = x * (y * z)$ obviously holds if two or more of the elements x,y,z are in S. Also if $z \in S$, then

$$b' * (b' * z) = b(bz) = (bb)z = (b' * b') * z,$$
$$b' * (c' * z) = b(cz) = (bc)z = (a\alpha)z = az = (b' * c') * z.$$

The other cases in which one of x,y,z is in S are treated similarly. Finally

$$b' * (c' * b') = b(cb) = (bc)b = (a\alpha)b = ab = (b' * c') * b',$$
$$c' * (c' * b') = c(cb) = (cc)b = (c' * c') * b',$$

and the other six cases are treated similarly. Consequently V is a strict extension of S by $Q = \{b,c,0\}$ with $Q^2 = 0$ By hypothesis, the extension V is determined by a partial homomorphism β so that $a = b' * c' = (b'\beta)(c'\beta)$. Now using the fact that $S^2 \subseteq R$ (since S is an inflation of R) we have $a = (b'\beta)(c'\beta) \in R$.

iii) *implies* iv). Suppose that $\pi_x = \pi_y$, $x \in S^2$. Then $x \in Z_a$, $y \in Z_b$ for some $a,b \in R$. It follows that $\pi_a = \pi_x = \pi_y = \pi_b$ in S and thus also $\pi_a = \pi_b$ in R. Since R is weakly reductive, we must have $a = b$. On the other hand, $x \in S^2$ implies that $Z_a = \{a\} = \{x\}$ since S is an inflation of R over $R \backslash R^2$. Thus $b = a = x \in S^2$ which implies that $Z_b = \{b\} = \{y\}$ since the inflation is over $R \backslash R^2$. Consequently $x = y$ as required.

iv *implies* ii) Let V be a strict extension of S and write $\tau = \tau(V:S)$. Let $\varphi: V \backslash S \to S$ be any function satisfying the condition $\tau^v = \pi_{v\varphi}$ for all $v \in V \backslash S$; its existence is assured by the axiom of choice. Hence $vs = (v\varphi)s$ and $sv = s(v\varphi)$ for all $s \in S$, $v \in V$. Next let $a,b \in V \backslash S$. If $ab \in S$, then

$$\pi_{ab} = \tau^{ab} = \tau^a\tau^b = \pi_{a\varphi}\pi_{b\varphi} = \pi_{(a\varphi)(b\varphi)}$$

so that $ab = (a\varphi)(b\varphi)$ since $(a\varphi)(b\varphi) \in S^2$. If $ab \notin S$, then

$$\pi_{(ab)\varphi} = \tau^{ab} = \tau^a\tau^b = \pi_{a\varphi}\pi_{b\varphi} = \pi_{(a\varphi)(b\varphi)}$$

so that $(ab)\varphi = (a\varphi)(b\varphi)$ as before. Consequently φ is a partial homomorphism and determines the extension.

ii) *implies* i). If V is a proper strict extension of S, then it is a retract extension and hence there is an S-homomorphism φ of

V. The congruence induced by φ is an S-congruence different from the equality relation, which proves that V is not a dense extension of S.

III.6.11 COROLLARY. If S is an inflation of a weakly reductive semigroup R, then $\Pi(S) \cong R$. In particular, if $\Pi(S)$ is globally idempotent, and strict and retract extensions of S coincide, then S is weakly reductive.

Proof. Exercise.

III.6.12. Exercises

1. With the notation of 6.2, prove the following statements.
   i) V is a strict extension if and only if $D = S$.
   ii) V is a pure extension if and only if $Q^*\varphi \cap S = \varnothing$.
   iii) V is a dense extension if and only if φ is one-to-one.
2. Find all semigroups of order ≤ 4 which are dense extensions of a zero semigroup of order 2.
3. Give an example of a semigroup which is a strict dense extension of some semigroup.
4. Give an example of a noncommutative semigroup which is a pure dense extension of a commutative semigroup S by a commutative semigroup Q.

III.6.13 REFERENCES: Grillet [1], Grillet and Grillet [1], Grillet and Petrich [1], Petrich and Grillet [1].

III.7. Semilattice Compositions

Using the extension theory developed in this chapter, we are now able to consider the problem opposite to that of semilattice decomposition of a semigroup, viz. that of semilattice composition. This problem can be stated thus: Given a semilattice Y and a collection of pairwise disjoint semigroups S_α indexed by Y, construct all semigroups S which admit a homomorphism φ onto Y and for which $\alpha\varphi^{-1} \cong S_\alpha$ for all $\alpha \in Y$. Stated differently, S can be taken to be the union of all S_α and must have a multiplication for which $S_\alpha S_\beta \subseteq S_{\alpha\beta}$ for all $\alpha,\beta \in Y$. (Caution: Such an S need not exist.) The following concept represents a ramification of a "semilattice of semigroups" defined in II.1.14.

III.7.1 DEFINITION. A semigroup S is a *semilattice Y of semigroups S_α* if there exists a homomorphism φ of S onto the semilattice Y such that $S_\alpha = \alpha\varphi^{-1}$ for all $\alpha \in Y$.

The next theorem gives a construction for the general case.

III.7.2 THEOREM. Let Y be a semilattice; for every $\alpha \in Y$ let S_α be a semigroup, D_α be an extension of S_α, and assume that $D_\alpha \cap D_\beta = \varnothing$ if $\alpha \neq \beta$. For every pair $\alpha,\beta \in Y$ such that $\alpha \geq \beta$, let $\psi_{\alpha,\beta}: S_\alpha \to D_\beta$ be a function satisfying:

i) $\psi_{\alpha,\alpha}$ is the identity mapping on S_α,

ii) $(S_\alpha\psi_{\alpha,\alpha\beta})(S_\beta\psi_{\beta,\alpha\beta}) \subseteq S_{\alpha\beta}$,

iii) if $\alpha\beta > \gamma$, then for all $a \in S_\alpha$, $b \in S_\beta$,

$$[(a\psi_{\alpha,\alpha\beta})(b\psi_{\beta,\alpha\beta})]\psi_{\alpha\beta,\gamma} = (a\psi_{\alpha,\gamma})(b\psi_{\beta,\gamma}).$$

On $S = \bigcup_{\alpha\in Y} S_\alpha$ define a multiplication $*$ by:

$$a * b = (a\psi_{\alpha,\alpha\beta})(b\psi_{\beta,\alpha\beta}) \qquad (a \in S_\alpha,\ b \in S_\beta). \tag{1}$$

Then S is a semilattice Y of semigroups S_α, in notation $S = (Y;S_\alpha,\psi_{\alpha,\beta},D_\alpha)$. Conversely, every semigroup S which is a semilattice Y of semigroups S_α can be so constructed. In addition, D_α can be chosen to satisfy:

iv) $D_\alpha = B_\alpha$, where $B_\alpha = \{b\psi_{\beta,\alpha} \mid b \in S_\beta,\ \beta \geq \alpha\}$,

v) D_α is a dense extension of S_α.

Furthermore, S is a subdirect product of semigroups B_α with a zero possibly adjoined.

Proof. Note that because of ii), $a * b \in S_{\alpha\beta}$ in (1), and that iii) holds also for $\alpha\beta = \gamma$ by i). Let S be as given above. Then for $a \in S_\alpha$, $b \in S_\beta$, $c \in S_\gamma$, we obtain

$$\begin{aligned}(a * b) * c &= [(a\psi_{\alpha,\alpha\beta})(b\psi_{\beta,\alpha\beta})] * c \\ &= [(a\psi_{\alpha,\alpha\beta})(b\psi_{\beta,\alpha\beta})]\psi_{\alpha\beta,\alpha\beta\gamma}(c\psi_{\gamma,\alpha\beta\gamma}) \\ &= (a\psi_{\alpha,\alpha\beta\gamma})(b\psi_{\beta,\alpha\beta\gamma})(c\psi_{\gamma,\alpha\beta\gamma})\end{aligned}$$

and similarly $a * (b * c)$ is equal to the same expression. Hence S is a semilattice Y of semigroups S_α.

Conversely, let S be a semilattice Y of semigroups S_α. For every $\alpha \in Y$, let $F_\alpha = \bigcup_{\beta \geq \alpha} S_\beta$. Then F_α is an extension of S_α and by 6.2 we can write $F_\alpha = [S_\alpha, F_\alpha/S_\alpha; \varphi_\alpha, D_\alpha]$ with all the conditions on D_α in 6.2. Now let $\psi_{\beta,\alpha} = \varphi_\alpha|_{S_\beta}$ for all $\beta > \alpha$, and let $\psi_{\alpha,\alpha}$ be the identity function on S_α. Then i), iv), and v) are satisfied. Next let $a \in S_\alpha$, $b \in S_\beta$. If $\alpha \leq \beta$, then using i), we obtain

$$(a\psi_{\alpha,\alpha\beta})(b\psi_{\beta,\alpha\beta}) = a(b\varphi_\alpha) \in S_\alpha = S_{\alpha\beta}$$

since S_α is an ideal of D_α; the case $\beta \leq \alpha$ is symmetric. If α and β are not comparable, then $a,b \in F_{\alpha\beta} \backslash S_{\alpha\beta}$ and $ab \in S_{\alpha\beta}$ so that

$$(a\psi_{\alpha,\alpha\beta})(b\psi_{\beta,\alpha\beta}) = (a\varphi_{\alpha\beta})(b\varphi_{\alpha\beta}) \in S_{\alpha\beta}$$

since $ab = 0$ in $F_{\alpha\beta}/S_{\alpha\beta}$. Hence ii) and ab $= (a\psi_{\alpha,\alpha\beta})(b\psi_{\beta,\alpha\beta})$ have been established.

Suppose that $\alpha\beta > \gamma$ and let $a \in S_\alpha$, $b \in S_\beta$. Then $ab \in S_{\alpha\beta}$ and

$$[(a\psi_{\alpha,\alpha\beta})(b\psi_{\beta,\alpha\beta})]\psi_{\alpha\beta,\gamma} = (ab)\varphi_\gamma = (a\varphi_\gamma)(b\varphi_\gamma) = (a\psi_{\alpha,\gamma})(b\psi_{\beta,\gamma})$$

proving iii).

For every $\alpha \in Y$, let $C_\alpha = B_\alpha$ if α is the zero of Y and $C_\alpha = B_\alpha \cup 0_\alpha$ otherwise, where 0_α acts as the zero. Further let θ_α be defined on S by

$$b\theta_\alpha = \begin{cases} b\psi_{\beta,\alpha} & \text{if} \quad b \in S_\beta, \quad \beta \geq \alpha, \\ 0_\alpha & \text{otherwise.} \end{cases}$$

For any $b \in S_\beta$, $c \in S_\gamma$, we obtain

$$\begin{aligned}(b\theta_\alpha)(c\theta_\alpha) &= (b\psi_{\beta,\alpha})(c\psi_{\gamma,\alpha}) = [(b\psi_{\beta,\beta\gamma})(c\psi_{\gamma,\beta\gamma})]\psi_{\beta\gamma,\alpha} \\ &= (bc)\psi_{\beta\gamma,\alpha} = (bc)\theta_\alpha\end{aligned}$$

if $\beta\gamma \geq \alpha$, and $(b\theta_\alpha)(c\theta_\alpha) = 0_\alpha = (bc)\theta_\alpha$ otherwise. Hence θ_α is a homomorphism of S onto C_α.

Now suppose that $b\theta_\alpha = c\theta_\alpha$ for all $\alpha \in Y$. Then $\beta \geq \alpha$ if and only if $\gamma \geq \alpha$ for all $\alpha \in Y$ which implies that $\beta = \gamma$. But then for $\alpha = \beta = \gamma$, we obtain $b = b\theta_\alpha = c\theta_\alpha = c$.

Therefore the mapping χ defined by:

$$\chi : a \to (a\theta_\alpha)_{\alpha \in Y} \qquad (a \in S)$$

is an isomorphism of S onto a subdirect product of $\{C_\alpha\}_{\alpha \in Y}$.

For $\alpha = \beta$, condition iii) implies that the function $\psi_{\alpha,\gamma}$ is a homomorphism. Condition ii) serves only to make condition iii) meaningful, and is automatically satisfied if Y is a chain. The theorem is stated for a fixed semilattice Y and given S_α; of course, we may let Y belong to special class $\mathcal{Y}$ of semilattices and S_α belong to a special class $\mathcal{C}$ of semigroups, in which case we obtain the class $\mathcal{F}$ of all semigroups S which are semilattices Y in $\mathcal{Y}$ of semigroups S_α in $\mathcal{C}$. We require here that both $\mathcal{Y}$ and $\mathcal{C}$ be closed under isomorphisms. If one is given a class $\mathcal{F}$ of semigroups, one takes for the class $\mathcal{C}$ the smallest possible class such that every semigroup in $\mathcal{F}$ is a semilattice of semigroups in $\mathcal{C}$. Another instance arises when one is given $\mathcal{Y}$ as the class of all chains and one studies the relationship between $\mathcal{C}$ and $\mathcal{F}$. These are the most frequent cases. Now taking the class $\mathcal{F}$ of all semigroups, we know that taking for $\mathcal{Y}$ the class of all semilattices, we may take for $\mathcal{C}$ the class of all $\mathfrak{N}$-simple semigroups by II.2.13. We thus have:

III.7.3 COROLLARY. Every semigroup is isomorphic to some $(Y;S_\alpha,\psi_{\alpha,\beta},D_\alpha)$ where S_α is $\mathfrak{N}$-simple, D_α is a dense extension of S_α and iv) in 7.2 holds.

III.7.4 COROLLARY. Every semigroup is a subdirect product of dense extensions of $\mathfrak{N}$-simple semigroups with a zero possibly adjoined.

We can modify the above construction by taking D_α to be an extension of an isomorphic copy of S_α. For example, if each S_α is weakly reductive, we may take $D_\alpha = \Omega(S_\alpha)$ which is a dense extension of $\Pi(S_\alpha)$ which is in turn an isomorphic copy of S_α. We thus have:

III.7.5 COROLLARY. Let Y be a semilattice; for every $\alpha \in Y$ let S_α be a weakly reductive semigroup, and assume that S_α are pairwise disjoint. For every pair $\alpha,\beta \in Y$ such that $\alpha \geq \beta$, let $\psi_{\alpha,\beta}:S_\alpha \to \Omega(S_\beta)$ be a function satisfying:

i) $\psi_{\alpha,\alpha}$ is the canonical homomorphism $S_\alpha \to \Omega(S_\alpha)$,

ii) $(S_\alpha\psi_{\alpha,\alpha\beta})(S_\beta\psi_{\beta,\alpha\beta}) \subseteq \Pi(S_{\alpha\beta})$,

iii) if $\alpha\beta > \gamma$, then for all $a \in S_\alpha$, $b \in S_\beta$,

$$[(a\psi_{\alpha,\alpha\beta})(b\psi_{\beta,\alpha\beta})]\psi^{-1}_{\alpha\beta,\alpha\beta}\psi_{\alpha\beta,\gamma} = (a\psi_{\alpha,\gamma})(b\psi_{\beta,\gamma}).$$

On $S = \bigcup_{\alpha \in Y} S_\alpha$ define a multiplication $*$ by:

$$a * b = [(a\psi_{\alpha,\alpha\beta})(b\psi_{\beta,\alpha\beta})]\psi^{-1}_{\alpha\beta,\alpha\beta}.$$

Then S is a semilattice Y of semigroups S_α, in notation $S = (Y;S_\alpha,\psi_{\alpha,\beta})$. Conversely, every semigroup S which is a semilattice Y of weakly reductive semigroups S_α can be so constructed.

III.7.6 COROLLARY. A semigroup S is separative if and only if S is a subdirect product of cancellative semigroups with a zero possibly adjoined.

Proof. Necessity. Let S be a separative semigroup. We know by II.6.4 that S is a semilattice of cancellative semigroups, so $S = (Y;S_\alpha,\psi_{\alpha,\beta},D_\alpha)$ where each S_α is cancellative and D_α is a dense extension of S_α. But then 5.15 implies that D_α is cancellative, and hence by 7.2, S is a subdirect product of cancellative semigroups with a zero possibly adjoined.

Sufficiency. Let S be a subdirect product of semigroups S_α, where S_α is either cancellative or S_α has a zero and S_α^* is a cancellative semigroup. Let (x_α), $(y_\alpha) \in S$ and suppose that $(x_\alpha)^2 = (x_\alpha)(y_\alpha)$ and $(y_\alpha)^2 = (y_\alpha)(x_\alpha)$. Then for every $\alpha \in Y$, we have $x_\alpha^2 = x_\alpha y_\alpha$ and $y_\alpha^2 = y_\alpha x_\alpha$. If S_α has no zero, we obtain $x_\alpha = y_\alpha$. If S_α has a zero, it follows that $x_\alpha = 0$ if and only if $y_\alpha = 0$; in case that $x_\alpha \neq 0$, we have $y_\alpha \neq 0$, so that $x_\alpha^2 = x_\alpha y_\alpha$ implies that $x_\alpha = y_\alpha$. Thus in any case, $x_\alpha = y_\alpha$ which yields $(x_\alpha) = (y_\alpha)$. One verifies similarly that $(x_\alpha)^2 = (y_\alpha)(x_\alpha)$ and $(y_\alpha)^2 = (x_\alpha)(y_\alpha)$ imply $(x_\alpha) = (y_\alpha)$. Therefore S is separative.

The next construction gives a special kind of semilattice of semigroups S_α; it is reminiscent of a retract extension.

III.7.7 PROPOSITION. Let Y be a semilattice; for every $\alpha \in Y$ let S_α be a semigroup and assume that the semigroups S_α are pairwise disjoint. For every pair $\alpha,\beta \in Y$ such that $\alpha \geq \beta$, let $\psi_{\alpha,\beta}:S_\alpha \to S_\beta$ be a homomorphism such that $\psi_{\alpha,\alpha}$ is the identity mapping and

$$\psi_{\alpha,\beta}\psi_{\beta,\gamma} = \psi_{\alpha,\gamma} \quad \text{if} \quad \alpha > \beta > \gamma.$$

Let $S = \bigcup_{\alpha \in Y} S_\alpha$ with multiplication:

$$a * b = (a\psi_{\alpha,\alpha\beta})(b\psi_{\beta,\alpha\beta}) \qquad (a \in S_\alpha,\ b \in S_\beta).$$

Then S is a semilattice Y of semigroups S_α, and is a subdirect product of semigroups S_α with a zero possibly adjoined.

Proof. Exercise.

The following concepts represent specializations of a semilattice of semigroups S_α.

III.7.8 DEFINITION. The semigroup S constructed in 7.7 is a *strong semilattice of semigroups* S_α and will be denoted by $[Y;S_\alpha,\psi_{\alpha,\beta}]$. The set of homomorphisms $\psi_{\alpha,\beta}$ satisfying the conditions of 7.7 is a *transitive system*. If, in addition, all $\psi_{\alpha,\beta}$ are one-to-one, S is a *sturdy semilattice of semigroups* S_α and will be denoted by $\langle Y;S_\alpha,\psi_{\alpha,\beta}\rangle$.

The following weakening of both left and right cancellativity will play an important role in the next chapter.

III.7.9 DEFINITION. A semigroup S is *weakly cancellative* if for any $a,b \in S$, $ax = bx$ and $xa = xb$ for some $x \in S$ imply $a = b$.

III.7.10 PROPOSITION. Let $S = (Y;S_\alpha,\psi_{\alpha,\beta},S_\alpha)$ and suppose that every S_α is weakly cancellative. Then S is a strong semilattice of semigroups S_α.

Proof. It suffices to show that the system $\{\psi_{\alpha,\beta}\}$ is transitive. Let $\alpha > \beta > \gamma$ and $a \in S_\alpha$, $b \in S_\beta$. Using 7.2, we obtain

$$(a\psi_{\alpha,\beta}\psi_{\beta,\gamma})(b\psi_{\beta,\gamma}) = [(a\psi_{\alpha,\beta})b]\psi_{\beta,\gamma} = (a\psi_{\alpha,\gamma})(b\psi_{\beta,\gamma})$$

and analogously $(b\psi_{\beta,\gamma})(a\psi_{\alpha,\beta}\psi_{\beta,\gamma}) = (b\psi_{\beta,\gamma})(a\psi_{\alpha,\gamma})$, which by weak cancellation in S_γ yields $a\psi_{\alpha,\beta}\psi_{\beta,\gamma} = a\psi_{\alpha,\gamma}$, as required.

The next result will prove a useful tool in constructing subdirect products.

III.7.11 THEOREM. On $S = \langle Y;S_\alpha,\psi_{\alpha,\beta}\rangle$ define a relation σ by:

$$a \,\sigma\, b \quad \text{if} \quad a\psi_{\alpha,\alpha\beta} = b\psi_{\beta,\alpha\beta} \qquad (a \in S_\alpha,\ b \in S_\beta).$$

Then σ is a congruence, S is a subdirect product of Y and S/σ, and if all S_α have any of the properties: (left, right, weak) cancellation or reductivity or (left, right, $\mathfrak{N}$-) simplicity, then S/σ has the same property.

Proof. It is clear that σ is reflexive and symmetric. Let $a\,\sigma\, b$, $b\,\sigma\, c$, $a \in S_\alpha$, $b \in S_\beta$, $c \in S_\gamma$, $\delta = \alpha\beta\gamma$. Then

$$\begin{aligned} a\psi_{\alpha,\delta} &= a\psi_{\alpha,\alpha\beta}\psi_{\alpha\beta,\delta} = b\psi_{\beta,\alpha\beta}\psi_{\alpha\beta,\delta} = b\psi_{\beta,\delta} \\ &= b\psi_{\beta,\beta\gamma}\psi_{\beta\gamma,\delta} = c\psi_{\gamma,\beta\gamma}\psi_{\beta\gamma,\delta} = c\psi_{\gamma,\delta} \end{aligned}$$

so that

$$a\psi_{\alpha,\alpha\gamma}\psi_{\alpha\gamma,\delta} = c\psi_{\gamma,\alpha\gamma}\psi_{\alpha\gamma,\delta}.$$

But then $a\psi_{\alpha,\alpha\gamma} = c\psi_{\gamma,\alpha\gamma}$ since $\psi_{\alpha\gamma,\delta}$ is one-to-one, proving transitivity.

Now suppose that $a\,\sigma\, b$ with $a \in S_\alpha$, $b \in S_\beta$, $c \in S_\gamma$, $\delta = \alpha\beta\gamma$. Then

$$\begin{aligned} (a * c)\psi_{\alpha\gamma,\delta} &= [(a\psi_{\alpha,\alpha\gamma})(c\psi_{\gamma,\alpha\gamma})]\psi_{\alpha\gamma,\delta} = (a\psi_{\alpha,\delta})(c\psi_{\gamma,\delta}) \\ &= (a\psi_{\alpha,\alpha\beta}\psi_{\alpha\beta,\delta})(c\psi_{\gamma,\delta}) \\ &= (b\psi_{\beta,\alpha\beta}\psi_{\alpha\beta,\delta})(c\psi_{\gamma,\delta}) = (b\psi_{\beta,\delta})(c\psi_{\gamma,\delta}) \\ &= (b\psi_{\beta,\beta\gamma}\psi_{\beta\gamma,\delta})(c\psi_{\gamma,\beta\gamma}\psi_{\beta\gamma,\delta}) \\ &= [(b\psi_{\beta,\beta\gamma})(c\psi_{\gamma,\beta\gamma})]\psi_{\beta\gamma,\delta} = (b * c)\psi_{\beta\gamma,\delta} \end{aligned}$$

and thus $a * c \;\sigma\; b * c$. Hence σ is a right congruence; one shows analogously that σ is also a left congruence.

Now define a mapping φ on S by

$$\varphi{:}a \rightarrow (\alpha, a\sigma) \qquad (a \in S_\alpha,\ \alpha \in Y).$$

If $a\varphi = b\varphi$, then $a,b \in S_\alpha$ for some α and $a = a\psi_{\alpha,\alpha} = b\psi_{\alpha,\alpha} = b$, so φ is an isomorphism of S into $Y \times S/\sigma$. It is clear that both projection homomorphisms of $S\varphi$ are onto, making S a subdirect product of Y and S/σ.

Suppose that each S_α is right cancellative, and let $(a\sigma)(c\sigma) = (b\sigma)(c\sigma)$. Then for $a \in S_\alpha$, $b \in S_\beta$, $c \in S_\gamma$, $\delta = \alpha\beta\gamma$, we have $a * c \;\sigma\; b * c$ so that $(a * c)\psi_{\alpha\gamma,\delta} = (b * c)\psi_{\beta\gamma,\delta}$ and thus

$$[(a\psi_{\alpha,\alpha\gamma})(c\psi_{\gamma,\alpha\gamma})]\psi_{\alpha\gamma,\delta} = [(b\psi_{\beta,\beta\gamma})(c\psi_{\gamma,\beta\gamma})]\psi_{\beta\gamma,\delta}.$$

Consequently

$$(a\psi_{\alpha,\delta})(c\psi_{\gamma,\delta}) = (b\psi_{\beta,\delta})(c\psi_{\gamma,\delta}),$$

which by right cancellation in S_δ yields $a\psi_{\alpha,\delta} = b\psi_{\beta,\delta}$. But then $a\psi_{\alpha,\alpha\beta}\psi_{\alpha\beta,\delta} = b\psi_{\beta,\alpha\beta}\psi_{\alpha\beta,\delta}$ which implies $a\psi_{\alpha,\alpha\beta} = b\psi_{\beta,\alpha\beta}$ since $\psi_{\alpha\beta,\delta}$ is one-to-one. Thus $a\ \sigma\ b$ and hence $a\sigma = b\sigma$, proving that S/σ is right cancellative.

Next assume that each S_α is simple and let $a \in S_\alpha$, $b \in S_\beta$. Then $a\psi_{\alpha,\alpha\beta}, b\psi_{\beta,\alpha\beta} \in S_{\alpha\beta}$ and hence $a\psi_{\alpha,\alpha\beta} = x(b\psi_{\beta,\alpha\beta})y$ for some $x,y \in S_{\alpha\beta}$ since $S_{\alpha\beta}$ is simple. But then $a\psi_{\alpha,\alpha\beta} = (x * b * y)\psi_{\alpha\beta,\alpha\beta}$ which implies that $a\ \sigma\ x * b * y$, and thus in S/σ, the equation $a\sigma = z(b\sigma)w$ has a solution which proves that S/σ is simple.

Suppose that each S_α is $\mathfrak{N}$-simple. Let I be a completely prime ideal of S/σ, and for every $\alpha \in Y$, let

$$I_\alpha = \{a \in S_\alpha | a\sigma \in I\}.$$

Then I_α is a completely prime ideal of S_α if it is nonempty. Hence either $I_\alpha = \varnothing$ or $I_\alpha = S_\alpha$. Since $I \neq \varnothing$, there exists $\alpha \in Y$ such that $I_\alpha = S_\alpha$. Then for any $\beta \in Y$, we must have $I_{\alpha\beta} = S_{\alpha\beta}$. But then $I_\beta = S_\beta$ since $b\psi_{\beta,\alpha\beta} = c$ for some $b \in S_\beta$, $c \in S_\gamma$ so that $b\sigma = c\sigma \in I$. Consequently $I = S/\sigma$ is $\mathfrak{N}$-simple by II.2.12.

The remaining cases have proofs analogous to the proof of one of the cases considered and are left as exercises.

III.7.12. Exercises

1. Show that the congruence σ defined in II.6.4 on a separative semigroup S is the greatest band congruence on S, all of whose classes are weakly cancellative.

2. Show that a semigroup S which is a semilattice of semigroups S_α, each of which has an identity, is a subdirect product of semigroups S_α with a zero possibly adjoined.

3. Prove that every semilattice of semigroups S_α each of which has an identity and no other idempotent is strong (e.g., this holds for every semilattice of groups).

4. Prove that every $\mathcal{H}$-class of a semigroup S is a group if and only if S is regular and a subdirect product of groups with a zero possibly adjoined.

5. Let Y be a chain; to every $\alpha \in Y$ associate a semigroup S_α, suppose that S_α are pairwise disjoint, and on $S = \bigcup_{\alpha \in Y} S_\alpha$ define a multiplication $*$ by

$$a * b = b * a = b \quad \text{if} \quad a \in S_\alpha,\ b \in S_\beta,\ \alpha > \beta,$$

and retain the multiplication within each S_α. Show that S is a semigroup and express it in terms of 7.2. (S is called a *successively annihilating band* of semigroups S_α.)

6. Characterize all semigroups S with the property that every subset of S consisting of 2 elements is a subsemigroup. Also characterize all semigroups S all of whose nonempty subsets are subsemigroups.

7. Show that for a family of pairwise disjoint semigroups S_α with identity indexed by a semilattice Y, a semilattice composition relative to Y always exists.

8. Let $Y = \{\alpha,\beta,\zeta\}$ be a Kronecker semigroup having ζ as its zero. Give examples of semigroups S_α, S_β, S_ζ such that an extension of S_ζ by S_α^0 and an extension of S_ζ by S_β^0 exist, but no composition relative to Y exists.

9. Prove that if every S_α in 7.11 is (left, right, weakly) cancellative, then σ is the least congruence on S for which S/σ has the respective property.

10. Show that a semigroup which is a semilattice of commutative weakly reductive semigroups is commutative.

11. Let $S = [Y;S_\alpha,\psi_{\alpha,\beta}]$ and $S' = [Y';S'_\alpha,\psi'_{\alpha,\beta}]$. Let ξ be an isomorphism of Y onto Y'; for every $\alpha \in Y$, let η_α be an isomorphism of S_α onto $S'_{\alpha\xi}$, and assume that for $\alpha > \beta$, the following diagram is commutative:

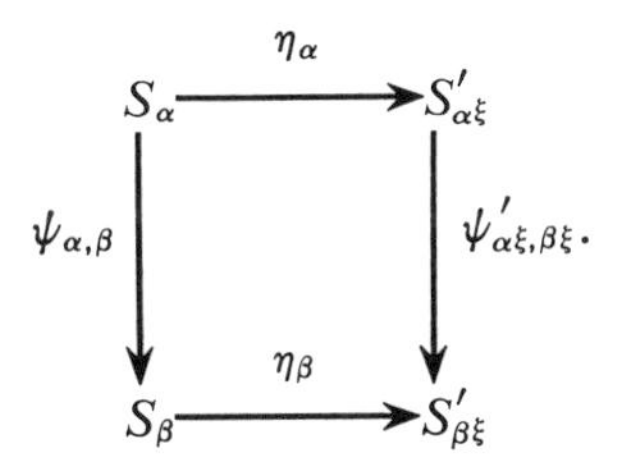

Show that the function χ defined on S by: $a\chi = a\eta_\alpha$ if $a \in S_\alpha$, is an isomorphism of S onto S'. Also show that if all S_α and S'_α are $\mathfrak{N}$-simple and weakly reductive, then every isomorphism of S onto S' can be so expressed.

III.7.13 REFERENCES: Arendt and Stuth [1], Brown and LaTorre [1], Clifford [5], Gantos [1], Kaufman [1], Lallement [1], O'Carroll and Schein [1], Preston [1], Schein [3], Skornjakov [1], Tamura [5], [16], [17], Yamada [2], [6], Yoshida and Yamada [1], Yoshida *et al.* [2].

IV

Completely Regular Semigroups

The class of completely regular semigroups coincides with the class of semigroups which are unions of their (maximal) subgroups. It is a subclass of the class of regular semigroups which we have already encountered. This subclass is particularly amenable to the treatment we have adopted in this book — the study of the structure of a semigroup through its greatest semilattice decomposition — for, as we will see, the $\mathcal{N}$-classes of such semigroups are completely simple, and the structure of the latter can be reduced to the structure of an arbitrary group. As far as the entire semigroup is concerned, one may use the results from the semilattice compositions, studied in III.7, of completely simple semigroups. In addition, we must know the translational hull of a completely simple semigroup, a task which will be undertaken in the next chapter.

After the general study of completely regular semigroups, we will characterize in several ways a completely simple semigroup. Then we will move on to various special classes of completely regular semigroups determined by different properties connected with their greatest semilattice decomposition and the properties of their idempotents.

IV.1. Generalities

In addition to the needed definitions, we will establish here several characterizations of the semigroups under study.

IV.1.1 DEFINITION. An element a of a semigroup S is *completely regular* if there exists $x \in S$ such that $a = axa$, $ax = xa$. A semigroup S is *completely regular* if all its elements are completely regular.

IV.1.2 PROPOSITION. The following conditions on an element a of a semigroup S are equivalent.

i) a is completely regular.
ii) a has an inverse with which it commutes.
iii) $a \in a^2Sa^2$.
iv) $a \in Sa^2 \cap a^2S$.
v) a is contained in a subgroup of S.

Proof. iv) *implies* v). If $a = xa^2 = a^2y$, then

$$xa = x(a^2y) = (xa^2)y = ay.$$

Letting $e = xa = ay$, we obtain

$$ae = a^2y = a = xa^2 = ea,$$
$$e^2 = (xa)(ay) = (xa^2)y = ay = e,$$
$$e \in Sa \cap aS,$$

which by I.4.11 implies $a \in G_e$.

The proof of the remaining implications (in circular order) is easy and is left as an exercise.

Part v) of 1.2 justifies the introduction of the concepts of completely regular elements and of completely regular semigroups, for the latter are precisely those semigroups which are unions of groups. (Hence these semigroups are also called "semigroups which are unions of groups.") For actual applications, part iv) usually turns out to be convenient.

IV.1.3 NOTATION. If a is a completely regular element of a semigroup S, we denote by a^{-1} the inverse of a in the maximal subgroup of S containing a.

IV.1.4 LEMMA. Let S be a semilattice of semigroups S_α, $\alpha \in Y$. Then S is (completely) regular if and only if S_α is (completely) regular for every $\alpha \in Y$. Furthermore, if $a \in S_\alpha$, then all inverses of a in S are contained in S_α.

Proof. It suffices to prove the last statement which is left as an exercise.

We will next characterize completely regular semigroups in several ways. For this, we need a lemma which is also of independent interest.

IV.1.5 LEMMA. Maximal subgroups of any semigroup coincide with the $\mathcal{H}$-classes containing idempotents. In particular, any two distinct maximal subgroups are disjoint.

Proof. This follows easily from I.4.11.

IV.1.6 THEOREM. The following conditions on a semigroup S are equivalent.

i) S is completely regular.
ii) For every $a \in S$, $a \in aSa^2$.
iii) S is a union of (disjoint) groups.
iv) Every $\mathcal{H}$-class is a group.
v) Every $\mathcal{N}$-class is simple and completely regular.
vi) Every left and every right ideal of S is completely semiprime.

Proof. i) *implies* ii). This follows from 1.2.

ii) *implies* iii). By 1.2, it suffices to show that for every $a \in S$, $a \in a^2S$. There exists $x \in S$ such that $a = axa^2$ and $y \in S$ such that $xa = (xa)y(xa)^2$. Consequently

$$\begin{aligned} a &= axa^2 = a(xa)a = a(xa)y(xa)^2a = (axayx)(axa^2) \\ &= axayxa = axay(xa) = (axay)(xa)y(xa)^2 \\ &= ay(xa)^2 = (ayx)a(xa) = (ayx)(axa^2)(xa) \\ &= (ay)(xa)^2(axa) = a(axa) = a^2xa \end{aligned}$$

so that $a \in a^2S$.

iii) *implies* iv). Let H be an $\mathcal{H}$-class. For any $a \in H$, a belongs to a maximal subgroup G_e. But then $a \mathcal{H} e$, so that $G_e = H$ by 1.5.

iv) *implies* v). Since $\mathcal{H} \subseteq \mathcal{N}$, each $\mathcal{N}$-class is a union of groups. By 1.4, each $\mathcal{N}$-class is completely regular. Also by 1.2, for every $a \in S$, $a = a^2xa^2$ for some $x \in S$ so that

$$a = a^2xa(a^2xa^2) \in Sa^2S,$$

which by II.4.5 implies that every $\mathcal{N}$-class is simple.

v) *implies* vi). If L is a left ideal of S and $a^2 \in L$, then since S is clearly completely regular, we have $a = xa^2$ by 1.2 for some $x \in S$, whence $a = xa^2 \in L$ and L is completely semiprime; the situation is analogous for right ideals.

vi) *implies* i). For any $a \in S$, Sa^2 is a left ideal and $a^4 \in Sa^2$. But then $a \in Sa^2$; similarly $a \in a^2S$, which by 1.2 implies that S is completely regular.

It is clear that in 1.6 we may substitute ii) by its left-right dual. Further, 1.6 says that every completely regular semigroup is a semilattice of *simple* completely regular semigroups; the next section is devoted to their precise structure.

A condition stronger than complete regularity is provided by the following.

IV.1.7 PROPOSITION. The following conditions on a semigroup S are equivalent.

i) S is a band of groups.
ii) S is completely regular and $\mathcal{H}$ is a congruence.
iii) S is regular and $a^2bS = abS$, $Sab^2 = Sab$ for all $a,b \in S$.

Proof. The equivalence of i) and ii) is a consequence of 1.6.

ii) *implies* iii). Since $a^2 \,\mathcal{H}\, a$ and $\mathcal{H}$ is a congruence, it follows that $a^2b \,\mathcal{H}\, ab$ so $a^2bS = abS$; similarly $Sab^2 = Sab$.

iii) *implies* ii). If $a = axa$, then $a^2xS = axS$ implies $a = axa \in a^2S$ and dually $a \in Sa^2$, and thus S is completely regular. Let $a \in G_e$ and $b \in S$. Then $a^2(a^{-1}b)S = a(a^{-1}b)S$ which implies $abS = ebS$. So if $a' \,\mathcal{H}\, a$, then $a'bS = ebS = abS$; also $Sa' = Sa$ implies $Sa'b = Sab$. Hence $a'b \,\mathcal{H}\, ab$ and $\mathcal{H}$ is a right congruence. By symmetry, we conclude that $\mathcal{H}$ is also a left congruence.

IV.1.8. Exercises

1. Prove that in a regular semigroup whose idempotents form a chain every $\mathcal{N}$-class is simple.

2. Show that each of the following conditions on a semigroup S is equivalent to regularity.

i) Every principal left ideal is generated by an idempotent.
ii) Every principal left ideal has a right identity.
iii) Every $\mathcal{L}$-class contains an idempotent.
(Note that because of the left-right symmetry of regularity, each of these statements is equivalent to its dual.)

3.* Prove that the following conditions on a regular semigroup S are equivalent.
i) E_S is a semilattice of left zero semigroups.
ii) If $a = axa = aya$, then $ax = ay$.
iii) Every $\mathcal{R}$-class contains a unique idempotent.

4.* Prove that the following conditions on a semigroup S are equivalent.
i) S is completely regular and Y_S is a chain.
ii) Every left and every right ideal is completely semiprime and every (two-sided) ideal is completely prime.
iii) For any $x,y \in S$, either $x \in xySx$ or $y \in yxSy$.

5. Show that in a regular semigroup S, any one of the conditions
i) $Sab = Sab^2$ for all $a,b \in S$,
ii) $\mathcal{L}$ is a congruence,
implies that S is completely regular and $\mathcal{H}$ is a left congruence.

6. Show that in a completely regular semigroup S, $\mathcal{K} \subseteq \mathcal{H}$ and the congruence generated by $\mathcal{H}$ is the least band congruence on S.

7. Show that every cyclic subsemigroup of a semigroup S has an identity if and only if S is completely regular and periodic.

8. Show that a regular semigroup whose idempotents are in the center is a semilattice of groups and conversely.

9. Show that a cancellative regular semigroup must be a group.

10. Give an example of a completely regular semigroup which is not a band of groups.

IV.1.8 REFERENCES: Clifford [1], [3], Croisot [1], Feller and Gantos [1], Petrich [2], [4].

IV.2. Completely Simple Semigroups

We have seen in the preceding section that a completely regular semigroup is a semilattice of completely regular simple semigroups and conversely. The purpose of this section is to give the precise structure of completely regular simple semigroups modulo groups, i.e., to reduce their structure to the structure of groups. In order to facilitate the proving of the main result as well as for didactic reasons, we first establish a theorem giving a number of characterizations of these semigroups. In particular, one characterization says that they are precisely all the completely simple semigroups. We then establish the principal result which says that every such semigroup is isomorphic to a Rees matrix semigroup over a group and conversely. This is a special case of the so-called Rees theorem for completely 0-simple semigroups which will not be discussed here.

IV.2.1 DEFINITION. An idempotent e of a semigroup S without zero is *primitive* if it is minimal relative to the partial order on E_S (i.e., $f^2 = f = ef = fe$ implies $f = e$). The unique element of a trivial semigroup is a primitive idempotent.

IV.2.2 LEMMA. A regular semigroup S all of whose idempotents are primitive is completely regular with maximal subgroups given by

$$G_e = aSa \qquad (e \in E_S,\ a \in G_e).$$

Proof. Let $a \in S$. Then $a = axa$ and $a^2 = a^2ya^2$ for some $x,y \in S$. Letting $e = ax$ and $f = a^2yax$, we obtain

$$\begin{aligned} f^2 &= (a^2yax)(a^2yax) = a^2y(axa)ayax = (a^2y)^2ax \\ &= (a^2y)ax = f, \\ ef &= (ax)(a^2yax) = (axa)(ayax) = a^2yax = f, \\ fe &= (a^2yax)(ax) = a^2yax = f, \end{aligned}$$

so that $e,f \in E_S$ and $e \geq f$. By hypothesis, we have $e = f$, whence $ax = a^2yax$, which, after multiplication by a on the right, yields $a = a^2ya \in a^2Sa$. Since a is arbitrary, 1.6 shows that S is completely regular.

Let $e \in E_S$ and $a \in G_e$. If $x \in G_e$, then

$$x = exe = (aa^{-1})x(a^{-1}a) \in aSa,$$

and thus $G_e \subseteq aSa$. Conversely, let $x = aya$. Then $x \in G_f$ for some $f \in E_S$ so that

$$ef = e(xx^{-1}) = e(aya)x^{-1} = (aya)x^{-1} = xx^{-1} = f$$

and dually $fe = f$ so that $e \geq f$. By hypothesis, we must have $e = f$ and hence $x \in G_e$. Since $x \in aSa$ is arbitrary, it follows that $aSa \subseteq G_e$.

IV.2.3 DEFINITION. A simple semigroup containing a primitive idempotent is *completely simple*.

We are now ready to prove the desired characterization theorem.

IV.2.4 THEOREM. The following conditions on a semigroup S are equivalent.

i) S is completely simple.
ii) S is completely regular and simple.
iii) S is regular and all its idempotents are primitive.
iv) S is regular and weakly cancellative.
v) S is regular and for any $a,x \in S$, $a = axa$ implies $x = xax$.

Proof. i) *implies* ii). Let e be a primitive idempotent of S and a be any element of S. Then by simplicity of S, we have $a = uev$ and $e = x(ea^3e)y$ for some $u,v,x,y \in S$. Letting $f = evaeyexeaue$, we obtain

$$\begin{aligned} f^2 &= evaeyexea(ueev)aeyexeaue = evaeye(xea^3ey)exeaue \\ &= evaeyexeaue = f \end{aligned}$$

and evidently $f \leq e$ so that $f = e$. Hence

$$a = ufv = (uev)aeyexea(uev) = a^2(eyexe)a^2 \in a^2Sa^2$$

and a is completely regular by 1.2.

ii) *implies* iii). Let $e,f \in E_S$ be such that $e \leq f$. Then by simplicity of $S, f = xey$ for some $x,y \in S$. Letting $a = fxf$ and $b = fyf$, we obtain

$$aeb = (fxf)e(fyf) = f(xey)f = f. \tag{1}$$

For a' satisfying the equalities $a = aa'a$, $aa' = a'a$, by (1) we have

$$f = aeb = aa'aeb = aa'f = a'af = a'a = aa' \tag{2}$$

and hence using (1) and (2), we obtain

$$\begin{aligned} f &= (a'a)(a'a) = a'(aa')a = a'fa = a'(aeb)a \\ &= (a'a)(eba) = feba = eba. \end{aligned}$$

But then $e = ef = f$ which implies that f is primitive.

iii) *implies* iv). Suppose that $ax = bx$ and $xa = xb$. By 2.2, S is completely regular, so $a \in G_e$ and $b \in G_f$ for some $e,f \in E_S$. Again by 2.2, $axa \in G_e$ and $bxb \in G_f$. Hence $axa = bxa = bxb$ implies that $e = f$. Letting $y = exe$, we see by 2.2 that $y \in G_e$, and also that

$$ay = a(exe) = (ae)xe = axe = bxe = (be)xe = b(exe) = by$$

which in G_e implies $a = b$, establishing that S is weakly cancellative.

iv) *implies* v). If $a = axa$, then $ax = a(xax)$ and $xa = (xax)a$, which by weak cancellation implies $x = xax$.

v) *implies* i). If $e,f \in E_S$ and $e \leq f$, then $e = efe$ so by hypothesis we also have $f = fef = e$ and every idempotent is primitive. By 2.2, S is completely regular. Suppose that S is not simple. Then by 1.6, S has more than one $\mathfrak{N}$-class, and in particular it must have two $\mathfrak{N}$-classes $N_e > N_f$ with $e,f \in E_S$. Letting $x \in eN_fe$, we have $x = eae \in G_g$ for some $a \in N_f$ and $g \in E_{N_f}$, whence

$$eg = exx^{-1} = e(eae)x^{-1} = (eae)x^{-1} = xx^{-1} = g$$

and dually $ge = g$. Hence $g < e$ since $g \in N_f$, contradicting the fact that all idempotents of S are primitive.

This theorem gives some useful characterizations of the semigroups under consideration. We turn next to their precise structure.

IV.2.5 LEMMA. Let S be a completely simple semigroup and let $e,f \in E_S$. Then the following statements hold.

i) For any $a,b \in S$, $ab \in G_e$ implies $aSb \subseteq G_e$.
ii) $ef = e$ implies $fe = f$.
iii) $ef = f$ implies $fe = e$.

Proof. Note that S satisfies the hypothesis of 2.2 in view of 2.4.

i) Let $a \in G_g$, $b \in G_h$, $ab \in G_e$. Then 2.2 implies $aba \in G_g$ and $bab \in G_h$. Hence $a = (aba)u$, $b = v(bab)$ for some $u \in G_g$, $v \in G_h$. For any $x \in S$, using 2.2, we obtain

$$axb = (aba)uxv(bab) = (ab)(auxvb)(ab) \in G_e$$

since $ab \in G_e$. Since x is arbitrary, we have $aSb \subseteq G_e$.

ii) If $ef = e$, then $fe \in E_S$ and also $fe = fef \in G_f$ by 2.2. But then $fe = f$.

iii) This is dual to ii).

IV.2.6 DEFINITION. Let G be a group, I and M be nonempty sets in which elements are denoted by $i,j,k,\ldots$ and $\mu,\nu,\epsilon,\ldots$, respectively, and let P:M $\times I \to G$ be any function; its value at (μ,i) is denoted by $p_{\mu i}$. Let S be the set $I \times G \times$ M together with the multiplication

$$(i,a,\mu)(j,b,\nu) = (i,ap_{\mu j}b,\nu).$$

Then S is a semigroup called the *Rees* $I \times$ M *matrix semigroup over the group* G *with the sandwich matrix* P and denoted by $S = \mathfrak{M}(I,G,\text{M};P)$.

We will usually call such a semigroup a *Rees matrix semigroup*. The sandwich matrix and the Rees matrix semigroup derive their names from the following "matrix" interpretation of the elements of this semigroup.

Let I and M be nonempty sets and G be a group; let 0 (called "zero") be any symbol not in G. Let S be the set consisting of all $I \times$ M matrices having precisely one entry from G. and the remaining entries 0. The fact that neither I nor M need be finite should cause no difficulty; one can easily imagine an array having an infinite number of "rows" and/or "columns" which has exactly one nonzero entry and this entry is taken from G. The matrix having its only nonzero entry $a \in G$ in the (i,μ) position is denoted by $(a)_{i\mu}$. We then think of P as an M $\times I$ matrix whose (μ,i)-entry is $p_{\mu i}$ and usually write $P = (p_{\mu i})$. We define a multiplication $*$ on the set of these matrices by: $A * B = APB$, where multiplication on the right is the "usual multiplication of matrices," i.e., row by column, where the summands containing 0 are considered as zero. Thus, for example, AP is an $I \times I$ matrix and APB is an $I \times$ M matrix having exactly one nonzero entry, so $A * B \in S$. The term "sandwich matrix" for P is thus justified. It is easy to see that the function $(i,a,\mu) \to (a)_{i\mu}$ is an isomorphism, i.e., the multiplication agrees with the one in 2.6.

An abstract semigroup T is said to have a *Rees matrix representation* $\mathfrak{M}(I,G,\text{M};P)$ if there exists an isomorphism of T onto $\mathfrak{M}(I,G,\text{M};P)$. This latter semigroup can be thought of as a semigroup of triples or of matrices. We are now ready to prove the structure theorem.

IV.2.7 THEOREM. Let S be a completely regular simple semigroup; fix $g \in E_S$, and let $G = G_g$,

$$I = \{e \in E_S \mid eg = e\}, \qquad \mathrm{M} = \{f \in E_S \mid gf = f\},$$

$P = (p_{fe})$ where $p_{fe} = fe$. Then the mapping χ defined by

$$\chi{:}a \to (e,gag,f) \qquad (a \in S)$$

where $ag \in G_e$, $ga \in G_f$, is an isomorphism of S onto $T = \mathfrak{M}(I,G,\mathrm{M};P)$.

Proof. 1. For any $f \in \mathrm{M}$, $e \in I$, we have

$$p_{fe} = fe = (gf)(eg) \in gSg = G_g$$

by 2.2 and 2.4 so that P is indeed a matrix over G. Also $gag \in G_g = G$ for any $a \in S$.

Let a be an element of S. If $ag \in G_e$, then $u(ag) = e$ for some $u \in G_e$, and we obtain

$$eg = u(ag)g = u(ag) = e,$$

which shows that $e \in I$. Similarly we see that $ga \in G_f$ implies $f \in \mathrm{M}$. Thus χ maps S into T.

2. Let $a,a' \in S$ with $ag \in G_e$, $ga \in G_f$, $a'g \in G_{e'}$, $ga' \in G_{f'}$; then by 2.5 i), $aa'g \in G_e$, $gaa' \in G_{f'}$, and we obtain

$$\begin{aligned}(a\chi)(a'\chi) &= (e,gag,f)(e',ga'g,f') = (e,(gag)(fe')(ga'g),f') \\ &= (e,(ga)(gf)(e'g)(a'g),f') = (e,[(ga)f][e'(a'g)],f') \\ &= (e,(ga)(a'g),f') = (e,g(aa')g,f') = (aa')\chi,\end{aligned}$$

and thus χ is a homomorphism.

3. With the same notation, suppose that $a\chi = a'\chi$. Then

$$(e,gag,f) = (e',ga'g,f')$$

so that $e = e'$, $gag = ga'g$, $f = f'$. Consequently

$$\begin{aligned}ga &= (ga)f = ga(gf) = (ga'g)f = (ga')(gf) \\ &= ga'(gf') = (ga')f' = ga'\end{aligned}$$

and dually $ag = a'g$. By 2.4, S is weakly cancellative, so $ag = a'g$ and $ga = ga'$ imply $a = a'$. Thus χ is one-to-one.

4. Let $(e,x,f) \in T$. Then $eg = e$ and $gf = f$, which by 2.5 ii) and iii) imply $ge = g = fg$.

Let $a = exf$. Then

$$e(ag)e = e(exf)ge = (exf)ge = ag$$

which by 2.5 i) implies $ag \in G_e$. Similarly we obtain $ga \in G_f$. Also

$$gag = g(exf)g = (ge)x(fg) = gxg = x.$$

Consequently $a\chi = (e,x,f)$ which proves that χ is onto.
Therefore χ is an isomorphism of S onto T.

We can sum up the structure theorem in the following form.

IV.2.8 COROLLARY. A semigroup S is completely simple if and only if S is isomorphic to a Rees matrix semigroup over a group.

Proof. Necessity follows from 2.7, while the proof of sufficiency consists of a simple calculation in a Rees matrix semigroup and is left as an exercise.

IV.2.9 PROPOSITION. The semigroup $S = \mathfrak{M}(I,G,\mathrm{M};P)$ is
i) the union of pairwise disjoint groups

$$H_{i\mu} = \{(i,a,\mu) \mid a \in G\} \qquad (i \in I,\ \mu \in \mathrm{M})$$

with the identity element $(i,p_{\mu i}^{-1},\mu)$,
ii) the union of pairwise disjoint minimal left ideals

$$L_\mu = \{(i,a,\mu) \mid i \in I,\ a \in G\} \qquad (\mu \in \mathrm{M}),$$

iii) the union of pairwise disjoint minimal right ideals

$$R_i = \{(i,a,\mu) \mid a \in G,\ \mu \in \mathrm{M}\} \qquad (i \in I).$$

Proof. Exercise.

IV.2.10. Exercises

1.* Prove that each of the following conditions on a semigroup S is equivalent to complete simplicity.

i) For every $a \in S$, the set aS contains an idempotent, and for any $a,x \in S$, $a = axa$ implies $x = xax$.

ii) For every $a \in S$, we have $a \in a^2S$, and for any $a,x \in S$, $a = a^2x$ implies $x = x^2a$.

iii) For any $a,b \in S$, $a \in abSa$.

2. Let φ be a homomorphism of a completely simple semigroup S into some semigroup T. Show that if φ is one-to-one both on idempotents and subgroups, then φ is one-to-one.

3. Show that every finite semigroup has a completely simple kernel.

4. Prove that a semigroup S is an inflation of a completely simple semigroup if and only if S^2 is completely simple and for any $x,y \in S$, $x^2 \in G_e$ and $y^2 \in G_f$ imply $xey = xfy = xy$.

IV.2.11 REFERENCES: Allen [1], Clifford [1], Lallement and Petrich [1], Rees [1], Steinfeld [1], Suškevič [1], Venkatesan [1].

IV.3. Semilattices of Rectangular Groups

We have seen in the two preceding sections that every completely regular semigroup is a semilattice of completely simple semigroups and conversely. In this section, we consider completely regular semigroups whose idempotents form a subsemigroup. It turns out that this condition is equivalent to the condition that the idempotents of each $\mathfrak{N}$-class form a subsemigroup. Each of these is completely simple, which coupled with the condition that the idempotents form a subsemigroup implies that they are rectangular groups. As special cases, we consider bands, semilattices of left groups and variations thereof.

We start with the following general result.

IV.3.1 PROPOSITION. The following three conditions on a semigroup S are equivalent.

i) E_S is a subsemigroup of S.

ii) For any $a,b \in S$ and their inverses a',b', the element $b'a'$ is an inverse of ab.

iii) For any $a,b,x,y \in S$, $a = axa$ and $b = byb$ imply $ab = abyxab$.

If S is a regular semigroup, then each of these conditions is equivalent to:

iv) Every inverse of every idempotent is idempotent.

Proof. i) *implies* ii). With the notation of ii), we have $a'a, bb' \in E_S$, and thus

$$(a'a)(bb')(a'a)(bb') = (a'a)(bb').$$

Multiplying this expression on the left by a and on the right by b, we obtain $abb'a'ab = ab$. Analogously, $a' = a'aa'$ and $b' = b'bb'$ imply $b'a'abb'a' = b'a'$, and hence $b'a'$ is an inverse of ab.

ii) *implies* iii). Let $a = axa$, $b = byb$. Then xax and yby are inverses of a and b, respectively, and thus $(yby)(xax)$ is an inverse of ab. Hence

$$ab = (ab)(yby)(xax)ab = abyxab.$$

iii) *implies* i). This follows without difficulty.

iii) *implies* iv). Let $e \in E_S$ and x be an inverse of e. Then

$$\begin{aligned} x = xex &= (xe)(ex) = [(xe)(ex)](ex)(xe)[(xe)(ex)] \\ &= (xex)^2 = x^2. \end{aligned}$$

Now suppose that S is regular.

iv) *implies* i). Let $e,f \in E_S$ and x be an inverse of ef. Then

$$\begin{aligned} fxe &= f(xefx)e = (fxe)^2, \\ (ef)(fxe)(ef) &= efxef = ef, \\ (fxe)(ef)(fxe) &= (fxe)^2 = fxe, \end{aligned}$$

so that ef is an inverse of the idempotent fxe. Thus by hypothesis we have $ef \in E_S$.

IV.3.2 COROLLARY. If S is a semilattice of regular semigroups S_α such that E_{S_α} is a subsemigroup of S_α, then E_S is a subsemigroup of S.

Proof. If $a \in S_\alpha$ and $x \in S_\beta$ is an inverse of a, then $a = axa$ implies $\alpha \leq \beta$ and $x = xax$ implies $\beta \leq \alpha$ and thus $\alpha = \beta$. Consequently if $e \in E_{S_\alpha}$ and x is an inverse of e, then $x \in S_\alpha$ and 3.1 implies that x is idempotent. But then 3.1 implies that E_S is a subsemigroup of S.

We consider now Rees matrix semigroups over a group in which idempotents form a subsemigroup. The next theorem gives several characterizations of such Rees matrix semigroups.

IV.3.3 THEOREM. The following conditions on $S = \mathfrak{M}(I,G,\mathrm{M};P)$ are equivalent.

i) Idempotents of S form a subsemigroup.
ii) For any $i,j \in I$, $\mu,\nu \in \mathrm{M}$, $p_{\mu i}^{-1}p_{\mu j}p_{\nu j}^{-1}p_{\nu i} = e$, the identity of G.
iii) There exist mappings $\alpha: I \to G$ and $\beta: \mathrm{M} \to G$ such that for any $i \in I$, $\mu \in \mathrm{M}$, $p_{\mu i} = (\mu\beta)(\alpha i)$.
iv) $S \cong I \times G \times \mathrm{M}$ where I and M are given the multiplication of a left and a right zero semigroup, respectively.

Proof. i) *implies* ii). According to 2.9, all idempotents of S are of the form $(i,p_{\mu i}^{-1},\mu)$, so the equality

$$(i,p_{\mu i}^{-1},\mu)(j,p_{\nu j}^{-1},\nu) = (i,p_{\mu i}^{-1}p_{\mu j}p_{\nu j}^{-1},\nu)$$

immediately implies $p_{\mu i}^{-1}p_{\mu j}p_{\nu j}^{-1} = p_{\nu i}^{-1}$ since the last element is in the $\mathcal{H}$-class $H_{i\nu}$.

ii) *implies* iii). Fix an element in each of the sets I and M and denote both of them by 1. Define α on I and β on M by

$$\alpha i = p_{11}^{-1}p_{1i}, \qquad \mu\beta = p_{\mu 1}.$$

Setting $\nu = j = 1$ in the identity in ii), we obtain $p_{\mu i} = p_{\mu 1}p_{11}^{-1}p_{1i} = (\mu\beta)(\alpha i)$.

iii) *implies* iv). Define a function $\chi: S \to I \times G \times \mathrm{M}$ by

$$\chi:(i,a,\mu) \to (i,(\alpha i)a(\mu\beta),\mu).$$

A straightforward verification that χ is an isomorphism of S onto $I \times G \times \mathrm{M}$ is left as an exercise.

iv) *implies* i). All idempotents of $I \times G \times \mathrm{M}$ are of the form (i,e,μ) where e is the identity of G. It is obvious that they form a subsemigroup, which by an isomorphism carries over to S.

The preceding result motivates the introduction of the following concept.

IV.3.4 DEFINITION. A semigroup isomorphic to the direct product of a rectangular band and a group is a *rectangular group*.

IV.3.5 COROLLARY. A regular semigroup whose idempotents form a rectangular band is a rectangular group and conversely. In particular, a rectangular band is the direct product of a left and a right zero semigroup and conversely.

Proof. Exercise.

IV.3.6 COROLLARY. A regular semigroup containing only one idempotent is a group.

Proof. Exercise.

IV.3.7 PROPOSITION. The following conditions on a semigroup S are equivalent.

i) Every $\mathfrak{N}$-class is a rectangular group.
ii) S is regular and $a = axa$ implies $a = ax^2a^2$.
iii) S is completely regular and E_S is a semigroup.

Proof. i) *implies* ii). Let $a = axa$. Then $ax \in N_{xa}$ so $(ax)(xa)(ax) = ax$ since $E_{N_{xa}}$ is a rectangular band. Hence $ax^2a^2 = (ax)(xa)(ax)a = axa = a$.

ii) *implies* iii). For any $a \in S$ there exists $x \in S$ such that $a = axa$ and thus also $a = ax^2a^2$. Consequently $a \in aSa^2$ and S is completely regular by 1.6. Let $e \in E_S$ and x be an inverse of e. Then $x = xex$ so by hypothesis $x = xex^2 = (xex)x = x^2$. Hence 3.1 implies that E_S is a subsemigroup of S.

iii) *implies* i). This follows from 1.6, 2.4, 2.8, and 3.3.

The following concepts represent special cases of a rectangular group.

IV.3.8 DEFINITION. A semigroup isomorphic to the direct product of a left (right) zero semigroup and a group is a *left* (*right*) *group*.

The next proposition gives a sample of a great number of abstract characterizations of left groups. Further characterizations can be found in the exercises. (One of these is: For any $a,b \in S$ there exists a unique $x \in S$ such that $xa = b$, which further justifies the name "left group.")

IV.3.9 THEOREM. The following conditions on a semigroup S are equivalent.

i) S is a left group.
ii) S is left simple and right cancellative.
iii) For every $a \in S$ there exists a unique $x \in S$ such that $xa^2 = a$.
iv) S is regular and E_S is a left zero semigroup.

Proof. i) *implies* ii). The proof consists of a simple verification and is left as an exercise.

ii) *implies* iii). Left simplicity implies solvability of $xa^2 = a$ while right cancellation insures the uniqueness of the solution.

iii) *implies* iv). Let $a,x \in S$ be such that $xa^2 = a$. Then

$$a = (xa)a = x(xa^2)a = (x^2a)a^2$$

which by hypothesis of uniqueness yields $x = x^2a$. For $y \in S$ satisfying $x = yx^2$, we similarly obtain $y = y^2x$, and also $xa = yx^2a = yx$. Using these equalities, we have

$$y = y^2x = y(yx) = y(xa) = (yx)a = (xa)a = xa^2 = a$$

which implies $xa = yx = ax$. This together with $a = xa^2$ implies that a is completely regular.

Next let $e,f \in E_S$ and $z \in S$ be such that $ef = z(ef)^2$. Then $ef = (ze)(ef)^2$ which by hypothesis yields $z = ze$. As above we also conclude that $z = z^2(ef) = z^2(ef)f = zf$. Consequently $ef = z(ef)^2 = z$ which together with $z = ze$ yields $ef = efe$. But then $ef = (ef)(ef)^2 = e(ef)^2$ and the hypothesis of uniqueness implies $ef = e$.

iv) *implies* i). This follows easily from 3.5.

IV.3.10 THEOREM. The following conditions on a semigroup S are equivalent.

i) Every $\mathfrak{N}$-class is a left group.
ii) S is regular and for any $a,x \in S$, $a = axa$ implies $ax = ax^2a$.
iii) S is completely regular and for any $e,f \in E_S$, $efe = ef$.
iv) S is regular and for every $a \in S$, $aS \subseteq Sa$.

Proof. i) *implies* ii). If $a = axa$, then $ax,xa \in E_{N_a}$ so that $ax = (ax)(xa) = ax^2a$ since E_{N_a} is a left zero semigroup.

ii) *implies* iii). For any $a \in S$ there is $x \in S$ such that $a = axa$. But then $ax = ax^2a$, whence

$$a = (ax)a = (ax^2a)a \in aSa^2$$

which by 1.6 implies that S is completely regular. Let $e,f \in E_S$. Then for some $u \in S$, we have $ef = efuef$ so that $ef = (ef)(fue)(ef)$ which by hypothesis implies

$$\begin{aligned} efue &= (ef)(fue) = ef(fue)^2ef = [(ef)(fue)(ef)](uef) \\ &= efuef = ef, \end{aligned}$$

so that

$$ef = efue = (efue)e = efe.$$

iii) *implies* iv). Let $a,b,x,y \in S$ be such that

$$a = axa, \qquad ax = xa, \qquad ab = abyab, \qquad aby = yab.$$

Then $ax,aby \in E_S$, and using the hypothesis, we obtain

$$\begin{aligned} ab &= ab(yab) = ab(aby) = ab(axa)by = (ab)(ax)(aby) \\ &= (ab)(ax)(aby)(ax) = (ababy)(xa) = abxa \end{aligned}$$

so that $ab \in Sa$ which proves that $aS \subseteq Sa$.

iv) *implies* i). For any $a \in S$, we have $a \in aSa \subseteq Sa^2$ which together with $aS \subseteq Sa$ by II.4.9 implies that every $\mathfrak{N}$-class is left simple. Let N be an $\mathfrak{N}$-class of S and $e,f \in E_N$. Then $Ne = Nf$ by left simplicity of N and hence $e = xf$ for some $x \in N$. But then $e = (xf)f = ef$ and thus N is a left group by 3.9.

IV.3.11 COROLLARY. The following conditions on a band B are equivalent.

i) Every $\mathfrak{N}$-class is a left zero semigroup.
ii) For any $e,f \in B$, $efe = ef$.
iii) For any $e \in B$, $eB \subseteq Be$.

Proof. Exercise.

IV.3.12. Exercises

1. Show that a band B has the property that for any $a,x \in B$, either $ax = axa$ or $xa = axa$ if and only if each $\mathfrak{N}$-class of B is either a

left or a right zero semigroup. Can the class of such bands be equationally defined?

2.* Prove that any one of the following conditions on a semigroup S is equivalent to S being a left group.
   i) S is left simple and contains an idempotent.
   ii) For any $a,b \in S$ there exists a unique $x \in S$ such that $xa = b$. (This is the original definition of a left group.)
   iii) S is regular and right cancellative.
   iv) S is regular and $a = axa$, $b = byb$ imply $a = ayb$.
   v) $P(S)$ is a group.

3. Prove that a semigroup S is an inflation of a rectangular group if and only if S^2 is a regular semigroup and for any $x,y \in S$ and $e \in E_S$, $xy = xey$.

4. Show that a semigroup S is a chain of rectangular groups if and only if S is completely regular and for any $e,f \in E_S$, either $e = efe$ or $f = fef$.

5. Show that a semigroup S is a chain of left groups if and only if S is completely regular and for any $e,f \in E_S$, either $e = ef$ or $f = fe$.

6. Show that a regular semigroup is a rectangular group if and only if for any $a,b,x,y \in S$, $a = axa$ and $b = byb$ imply $a = abyxa$.

7. Characterize all regular semigroups having at most two idempotents.

8. Let S be a semilattice of rectangular groups. Define a relation μ on S by:

$$a \,\mu\, b \quad \text{if for every } e \in E_S,\ a^{-1}ea = b^{-1}eb.$$

Show that μ is the greatest congruence on S contained in $\mathcal{H}$.

9.* Prove that every finitely generated subsemigroup of a semigroup S has a left identity if and only if S is a chain of periodic right groups. Derive the conditions in order that every finitely generated subsemigroup of S has an identity.

IV.3.13 REFERENCES: Chrislock [1], Clifford [5], Croisot [1], Fantham [1], T.E. Hall [1], [2], Ivan [1], Petrich [4], Yamada [5], [7], [9].

IV.4. Strong Semilattices of Completely Simple Semigroups

We have seen in the preceding chapter that among semilattice compositions the easiest ones to construct are the strong compositions; we called the semigroup in question "a strong semilattice of semigroups S_α." It is then natural, when considering completely regular semigroups, to study strong semilattices of completely simple semigroups. In view of the results of the present chapter, the problem of finding the structure of these semigroups reduces to finding transitive systems of homomorphisms among completely simple semigroups associated with elements of a semilattice.

We start with a lemma which is also of independent interest.

IV.4.1 LEMMA. Let S be a regular subsemigroup of a semigroup T. Then Green's relations $\mathcal{L}$, $\mathcal{R}$, $\mathcal{H}$ on S are the restrictions of those on T.

Proof. It suffices to consider $\mathcal{L}$; let $\mathcal{L}_S$ and $\mathcal{L}_T$ denote the corresponding $\mathcal{L}$-relations. Let $a,b \in S$ and $a \mathcal{L}_T b$. Then $a = ub$ and $b = va$ for some $u,v \in T^1$; letting a' and b' be inverses in S of a and b, respectively, we obtain $a = (ub)b'b = ab'b$ and $b = (va)a'a = ba'a$ which shows that $a \mathcal{L}_S b$. Consequently $\mathcal{L}_T|_S \subseteq \mathcal{L}_S$, the opposite inclusion is trivial.

We now investigate other varieties of bands.

IV.4.2 DEFINITION. A band is (*left,right*) *normal* if it satisfies the identity ($axy = ayx$, $xya = yxa$) $axya = ayxa$.

Note that left zero semigroups are left normal and that both rectangular bands and semilattices are normal. Recall that if a semigroup S is a band of groups, then S is completely regular and $\mathcal{H}$ is a congruence, and conversely; in such a case $S/\mathcal{H}$ is a band. Hence the statement "S is a normal band of groups" signifies that in addition $S/\mathcal{H}$ is a normal band. We are now ready to prove the principal result of this section.

IV.4.3 THEOREM. The following conditions on a semigroup S are equivalent.

i) S is a strong semilattice of completely simple semigroups.
ii) S is regular and a subdirect product of completely simple semigroups with a zero possibly adjoined.
iii) S is a normal band of groups.

iv) S is completely regular and for any $x \in S, f \in E_S$ such that $N_x < N_f$, there exists a unique $e \in E_{N_x}$ such that $e < f$.

Proof. i) *implies* ii). This follows immediately from III.7.7.

ii) *implies* iii). We may suppose that $S \subseteq \Pi_{\alpha \in A} S_\alpha$ where S is a subdirect product and each S_α is completely simple with a zero possibly adjoined. It is clear that in $T = \Pi_{\alpha \in A} S_\alpha$, $(a_\alpha) \mathcal{H} (b_\alpha)$ if and only if for all $\alpha \in A$, $a_\alpha \mathcal{H} b_\alpha$. It then follows from 4.1 that the same holds in S relative to each component S_α. A simple verification shows that $\mathcal{H}$ is a congruence on each S_α and hence $\mathcal{H}$ is a congruence on S. Further, for any $(a_\alpha) \in S$ and every $\alpha \in A$, $a_\alpha \mathcal{H} a_\alpha^2$ so that in S, $(a_\alpha) \mathcal{H} (a_\alpha)^2$. But then 1.2 quickly implies that (a_α) is completely regular. Furthermore, for each $\alpha \in A$, one verifies without difficulty that $S_\alpha/\mathcal{H}$ is a rectangular band with a zero eventually adjoined and thus $S_\alpha/\mathcal{H}$ is a normal band. Consequently for any $(a_\alpha),(x_\alpha),(y_\alpha) \in S$, we obtain for every $\alpha \in A$, $a_\alpha x_\alpha y_\alpha a_\alpha \mathcal{H} a_\alpha y_\alpha x_\alpha a_\alpha$ which then implies in S, $(a_\alpha)(x_\alpha)(y_\alpha)(a_\alpha) \mathcal{H} (a_\alpha)(y_\alpha)(x_\alpha)(a_\alpha)$ proving that $S/\mathcal{H}$ is a normal band.

iii) *implies* iv). Let $N_x < N_f$ where $f \in E_S$, and set $T = fN_xf$. For $y = fzf$ where $z \in N_x$, let y' be an inverse of y and let $w = fy'f$. One verifies without difficulty that w is an inverse of y. Since $y \in T \subseteq N_x$, we also have $y' \in N_x$ and thus $w \in T$. It follows that T is a regular subsemigroup of N_x and hence contains an idempotent, say e. But then clearly $e < f$ and $e \in N_x$. Suppose that also $g \in E_{N_x}$ and $g < f$. Letting $x \to \bar{x}$ be the natural homomorphism of S onto $S/\mathcal{H}$, we obtain $\bar{e} < \bar{f}$ and $\bar{g} < \bar{f}$, and, in addition, $\bar{e} \mathfrak{N} \bar{g}$. Since $S/\mathcal{H}$ is a normal band, we obtain

$$\bar{e} = \bar{e}\bar{g}\bar{e} = \bar{f}\bar{e}\bar{g}\bar{e}\bar{f} = \bar{f}\bar{g}\bar{e}\bar{g}\bar{f} = \bar{f}\bar{g}\bar{f} = \bar{g}.$$

But the natural homomorphism $S \to S/\mathcal{H}$ is clearly one-to-one on idempotents which implies $e = g$ as required.

iv) *implies* i). In view of III.4.7, for every $x \in S$, $N(x)$ is a retract extension of N_x. Hence using the notation of III.7.2, we may write $S \cong (Y;S_\alpha,\psi_{\alpha,\beta},S_\alpha)$ where $Y = Y_S$ and the S_α are the $\mathfrak{N}$-classes of S. According to 2.4, each S_α is weakly cancellative, which by III.7.10 implies that S is a strong semilattice of its $\mathfrak{N}$-classes. We also know by 2.4 that each $\mathfrak{N}$-class of S is completely simple.

IV.4.4 LEMMA. Let $S = L \times G \times R$, $S' = L' \times G' \times R'$ be two rectangular groups. For given functions $\varphi: L \to L'$, $\tau: R \to R'$; and a homomorphism $\omega: G \to G'$, let

$$(\ell,g,r)\theta = (\ell\varphi,g\omega,r\tau) \qquad ((\ell,g,r) \in S).$$

Then θ is a homomorphism of S into S'. Conversely, every homomorphism of S into S' can be so constructed.

Proof. The first statement is obvious. For the second, let $\theta : S \to S'$ be a homomorphism. Fix $\ell_0 \in L$, $r_0 \in R$, and define the functions φ,ω,τ by

$$(\ell,e,r_0)\theta = (\ell\varphi,e',r_0'),$$
$$(\ell_0,g,r_0)\theta = (\ell_0'',g\omega,r_0''),$$
$$(\ell_0,e,r)\theta = (\ell_0',e',r\tau).$$

Then $\varphi:L \to L'$, $\tau:R \to R'$, and $\omega:G \to G'$ is a homomorphism. Further

$$(\ell,g,r)\theta = (\ell,e,r_0)\theta(\ell_0,g,r_0)\theta(r_0,e,r)\theta = (\ell\varphi,g\omega,r\tau)$$

as required.

We will write $\theta = (\varphi,\omega,\tau)$.

We now introduce a special subdirect product.

IV.4.5 DEFINITION. Let $\{S_\alpha\}_{\alpha \in A}$ be a family of semigroups and assume that there exists a semilattice Y and for each $\alpha \in A$ an isomorphism φ_α of $S_\alpha/\mathfrak{N}$ onto Y. The set

$$S = \{(a_\alpha) \in \prod_{\alpha \in A} S_\alpha \mid N_{a_\alpha}\varphi_\alpha = N_{a_\beta}\varphi_\beta \quad \text{for all} \quad \alpha,\beta \in A\}$$

is a subsemigroup of $\prod_{\alpha \in A} S_\alpha$ called a *spined product* of the semigroups $\{S_\alpha\}_{\alpha \in A}$ (more precisely "relative to isomorphisms $\{\varphi_\alpha\}_{\alpha \in A}$").

We can visualize a spined product as that part of the direct product which consists of the "spine," e.g., for $|A| = 2$, we have an isomorphism $\varphi:S_1/\mathfrak{N} \to S_2/\mathfrak{N}$ and take $\bigcup_{a \in S_1}(N_a \times N_a\varphi)$.

IV.4.6 COROLLARY. The following conditions on a semigroup S are equivalent.

i) S is a strong semilattice of rectangular groups.
ii) S is regular and a subdirect product of semigroups of the following types: groups and left or right zero semigroups, and any of these with a zero adjoined.

iii) S is completely regular and E_S is a normal band.
iv) S is a spined product of a left normal band, a right normal band, and a semilattice of groups.

Proof. i) *implies* ii). By 4.3, S is a subdirect product of completely simple semigroups with a zero possibly adjoined, and the proof of 4.3 shows that these completely simple semigroups can be taken to be the $\mathfrak{N}$-classes of S. In the present case, the $\mathfrak{N}$-classes are rectangular groups. Further $(L \times G \times R)^0$ is a subdirect product of $L^0 \times G^0 \times R^0$ which, combined with the preceding statements, proves ii).

ii) *implies* iii). By 4.3, S is a normal band of groups. Since each of the semigroups T in ii) has the property that E_T is a subsemigroup of T, the same holds for S forcing E_S to be a normal band. (Equivalently, idempotents in each T form a normal band which carries over to any subsemigroup of their direct product.)

iii) *implies* iv). It follows that S is a semilattice of rectangular groups $L_\alpha \times G_\alpha \times R_\alpha$. Since E_S is a normal band, it satisfies iv) of 4.3, and hence also S satisfies iv) of 4.3. Consequently 4.3 implies that S is a strong semilattice of rectangular groups. Thus $S \cong [Y; L_\alpha \times G_\alpha \times R_\alpha, \psi_{\alpha,\beta}]$, and by 4.4, we have $\psi_{\alpha,\beta} = (\varphi_{\alpha,\beta},\omega_{\alpha,\beta},\tau_{\alpha,\beta})$ and the multiplication is given by: for $(\ell,g,r) \in S_\alpha$, $(\ell', g', r') \in S_{\alpha'}$,

$$(\ell,g,r)(\ell',g',r') = ((\ell\varphi_{\alpha,\alpha\alpha'})(\ell'\varphi_{\alpha',\alpha\alpha'}),(g\omega_{\alpha,\alpha\alpha'})(g'\omega_{\alpha',\alpha\alpha'}),(r\tau_{\alpha,\alpha\alpha'})(r'\tau_{\alpha',\alpha\alpha'}))$$

which evidently splits the multiplication into three parts each of which is a strong semilattice of its $\mathfrak{N}$-classes. The first part is easily seen to be a left normal band, the second is obviously a semilattice of groups, the third is a right normal band by considerations dual to those for the first part. Hence S is a spined product of these three semigroups.

iv) *implies* i). This follows easily from 4.4 and the definition of a spined product.

IV.4.7 COROLLARY. The following conditions on a band B are equivalent.

i) B is a strong semilattice of rectangular bands.
ii) B is a subdirect product of left and right zero semigroups with a zero possibly adjoined.

iii) B is a normal band.
iv) B is a spined product of a left and of a right normal band.

Proof. Exercise.

IV.4.8 COROLLARY. The following conditions on a band are equivalent.
i) B is a strong semilattice of left zero semigroups.
ii) B is a subdirect product of left zero semigroups with a zero possibly adjoined.
iii) B is a left normal band.

Proof. Exercise.

IV.4.9. Exercises

1.* Prove that each of the following conditions on a completely regular semigroup S is equivalent to any of the conditions in 4.3.
   i) For any $x,y \in S$ such that $N_x < N_y$, there exists $f \in E_{N_y}$ for which there is a unique $e \in E_{N_x}$ with the property $e < f$.
   ii) For any $a,x,y,b \in S$, $axyb \in aySxb$.
   iii) Both $\mathcal{L}$ and $\mathcal{R}$ are congruences, $S/\mathcal{L}$ is a right normal band, $S/\mathcal{R}$ is a left normal band.
   iv) Denoting by a^* the identity of the maximal subgroup of S containing a, we have for any $e,f,g \in E_S$ that $e \leq f$ implies $(eg)^* \leq (fg)^*$ and $(ge)^* \leq (gf)^*$.

2. Show that the following conditions on a semigroup S are equivalent.
   i) S is a spined product of a left normal band and a semilattice of groups.
   ii) S is a strong semilattice of left groups.
   iii) S is regular and for any $a,x,y \in S$, $axy \in aySx$.

3. Prove the following statements for a semigroup S which is a strong semilattice of semigroups S_α. If $\mathcal{H}$ is a congruence on each S_α, then the relation τ defined on S by: $a \,\tau\, b$ if $a,b \in S_\alpha$ for some

S_α and $a\mathcal{H}b$ in S_α, is a congruence. If also for every S_α, each element of S_α has a left and a right identity in S_α, then τ coincides with $\mathcal{H}$ on S. Deduce that in a strong semilattice of completely simple semigroups $\mathcal{H}$ is always a congruence.

4.* Call a band *left regular* if it satisfies the identity $xyx = xy$. Prove that the following conditions on a semigroup S are equivalent.
   i) $\mathcal{H}$ is a congruence and $S/\mathcal{H}$ is a left regular band.
   ii) S is regular and for any $a,b \in S$, $aS \subseteq Sa$ and $abS = a^2bS$.
   iii) S is a spined product of a left regular band and a semilattice of groups.

5. Derive an explicit construction of completely regular semigroups whose idempotents form a normal band.

6. Give an example of a semigroup which is a strong semilattice of completely simple semigroups but is neither completely simple nor a band nor a semilattice of groups.

7. Give an example of a semigroup which is a band of groups but is neither a strong semilattice of completely simple semigroups nor a band.

IV.4.10 REFERENCES: Howie [1], Kimura [1], Lallement [1], Petrich [6], Schein [1].

IV.5. Subdirect Products of a Semilattice and a Completely Simple Semigroup

Since every completely regular semigroup S is a semilattice of completely simple semigroups, it is natural to ask: Under which conditions is S a subdirect product of a semilattice and a completely simple semigroup? It turns out that in such a case the semilattice and the completely simple semigroup are respectively the greatest semilattice and the greatest completely simple homomorphic images of S. Intuitively speaking, we keep the same semilattice and find a completely simple semigroup which is an "abstraction" of all $\mathcal{N}$-classes of S. We then give a simple construction of all such semigroups within a direct product of a semilattice and a completely simple semigroup.

IV.5.1 THEOREM. A semigroup S is a sturdy semilattice of completely simple semigroups if and only if S is a regular semigroup subdirect product of a semilattice and a completely simple semigroup.

Proof. *Necessity.* By hypothesis we may take $S = \langle Y; S_\alpha, \psi_{\alpha,\beta} \rangle$ where each S_α is completely simple. In the notation of and according to III.7.11, S is a subdirect product of Y and S/σ. By 2.4, each S_α is weakly cancellative, so again by III.7.11, we have that S/σ is also. Since S is regular, so is S/σ which again by 2.4 shows that S/σ is completely simple.

Sufficiency. Let $S \subseteq Y \times T$ be a subdirect product where S is regular, Y is a semilattice, and T is completely simple. For every $\alpha \in Y$, let

$$S_\alpha = S \cap (\{\alpha\} \times T), \qquad T_\alpha = \{a \in T \mid (\alpha,a) \in S\}.$$

It is clear that S is a semilattice of semigroups S_α and that $S_\alpha \cong T_\alpha$. If (β,b) is an inverse in S of $(\alpha,a) \in S$, then it follows immediately that $\beta = \alpha$ and hence S_α is regular. But then T_α is a regular subsemigroup of the completely simple semigroup T which by 2.4 implies that T_α is completely simple. Consequently each S_α is completely simple.

Let $(\alpha,e),(\beta,f),(\beta,g) \in E_S$ and suppose that $(\alpha,e) > (\beta,f)$ and $(\alpha,e) > (\beta,g)$. Then $e \geq f$ and $e \geq g$ which in a completely simple semigroup implies $f = g = e$ so that $(\beta,f) = (\beta,g)$. Hence by 4.3 we conclude that S is a strong semilattice of semigroups S_α.

Let $(\alpha,a), (\alpha,b) \in S$, $\beta < \alpha$, and suppose that $(\alpha,a)\psi_{\alpha,\beta} = (\alpha,b)\psi_{\alpha,\beta}$. Then for any $(\beta,x) \in S$, we obtain.

$$(\alpha,a)(\beta,x) = [(\alpha,a)\psi_{\alpha,\beta}](\beta,x) = [(\alpha,b)\psi_{\alpha,\beta}](\beta,x) = (\alpha,b)(\beta,x)$$

which implies $ax = bx$, and analogously $xa = xb$. Consequently 2.4 implies that $a = b$ and thus $(\alpha,a) = (\alpha,b)$ proving that $\psi_{\alpha,\beta}$ is one-to-one. Therefore S is a sturdy semilattice of semigroups S_α.

IV.5.2 COROLLARY. A semigroup S is a sturdy semilattice of rectangular groups if and only if S is a regular semigroup subdirect product of a semilattice, a left zero semigroup, a group, and a right zero semigroup.

Proof. Necessity. According to 5.1, S is a subdirect product of a semilattice Y and a completely simple semigroup T. By 3.7, E_S is a subsemigroup of S, and since T is a homomorphic image of S, it follows that E_T is a subsemigroup of T. But then 3.3 implies that T is a rectangular group. Consequently S is a subdirect product of Y and $L \times G \times R$ where L, G and R are a left zero semigroup, a group, and a right zero semigroup, respectively. It follows immediately that then S is also a subdirect product of Y, L, G and R.

Sufficiency. By virtue of 5.1, S is a sturdy semilattice of completely simple semigroups S_α. Since the idempotents of each of the semigroups listed form a subsemigroup, the same holds for their direct product and thus also for S. But then 3.3 implies that each S_α is a rectangular group.

One deduces from 5.2 without difficulty characterizations of sturdy semilattices of (i) (left, right) groups, (ii) left (or right) zero semigroups, (iii) rectangular bands. The formulation and proofs of these statements are left as an exercise. We now turn to a construction of semigroups S in 5.1.

IV.5.3 COROLLARY. Let Y be a semilattice, T be a completely simple semigroup, $\mathcal{R}(T)$ be the partially ordered set of all regular subsemigroups of T ordered under inclusion, $\varphi: Y \to \mathcal{R}(T)$ be an order inverting function for which $\bigcup_{\alpha \in Y} \alpha\varphi = T$, and set

$$S = \{(\alpha,a) \in Y \times T \mid a \in \alpha\varphi\}.$$

Then S is a regular semigroup subdirect product of Y and T, and conversely, every regular semigroup subdirect product of Y and T and contained in $Y \times T$ can be so constructed.

Proof. The proof of the direct part consists of a straightforward verification and is left as an exercise. For the converse, we let $\alpha\varphi = T_\alpha$ for every $\alpha \in Y$ with the notation as in the proof of sufficiency of 5.1. It is easy to see that φ is inclusion inverting; the remaining properties have been established in the proof of sufficiency of 5.1.

For the semigroups S appearing in 5.2, we can give a more detailed construction. To this end, we need a lemma which is also of independent interest.

IV.5.4 LEMMA. For L, G and R a left zero semigroup, a group, and a right zero semigroup, respectively, the direct product $L \times G \times R$ is the only subdirect product of L, G and R contained in $L \times G \times R$.

Proof. Let $S \subseteq L \times G \times R$ be a subdirect product and let $(\ell,g,r) \in L \times G \times R$. Then $(\ell,h,r') \in S$ for some $h \in G$ and $r' \in R$; further $(u,h^{-1},v) \in S$ for some $u \in L$ and $v \in R$. Letting 1 denote the identity of G, we obtain

$$(\ell,1,r') = (\ell,h,r')(u,h^{-1},v)^2(\ell,h,r') \in S.$$

Analogously $(\ell',1,r) \in S$ for some $\ell' \in L$, also $(\ell'',g,r'') \in S$ for some $\ell'' \in L$ and $r'' \in R$, which yields

$$(\ell,g,r) = (\ell,1,r')(\ell'',g,r'')(\ell',1,r) \in S$$

as desired.

IV.5.5 THEOREM. Let L, G and R be a left zero semigroup, a group and a right zero semigroup; further let $\mathcal{P}(L)$ and $\mathcal{P}(R)$ denote the partially ordered sets of all nonempty subsets of L and R, respectively, and let $\mathcal{L}(G)$ denote the lattice of all subgroups of G. Finally let Y be a semilattice and let σ, ω and τ be order inverting functions as follows:

$$\begin{array}{ccccc} \mathcal{P}(L) & \xleftarrow{\sigma} & Y & \xrightarrow{\tau} & \mathcal{P}(R) \\ & & \downarrow{\scriptstyle\omega} & & \\ & & \mathcal{L}(G) & & \end{array}$$

satisfying

$$L = \bigcup_{\alpha \in Y} \alpha\sigma, \qquad G = \bigcup_{\alpha \in Y} \alpha\omega, \qquad R = \bigcup_{\alpha \in Y} \alpha\tau,$$

and set

$$S = \{(\alpha,\ell,g,r) \in Y \times L \times G \times R \mid \ell \in \alpha\sigma,\ g \in \alpha\omega,\ r \in \alpha\tau\}$$

with the coordinatewise multiplication. Then S is a regular semigroup subdirect product of Y, L, G and R. Conversely, every regular semigroup subdirect product of Y, L, G and R and contained in $Y \times L \times G \times R$ can be so constructed.

Proof. Again the proof of the direct part consists of a simple verification and is left as an exercise. For the converse, we let $S \subseteq Y \times L \times G \times R$ be a regular semigroup subdirect product of Y, L, G and R. It is easy to see that

$$T = \{(\ell,g,r) \in L \times G \times R \mid (\alpha,\ell,g,r) \in S \text{ for some } \alpha \in Y\}$$

is a subdirect product of L, G and R, so by 5.4 we must have $T = L \times G \times R$. It follows that S is a subdirect product of the semilattice Y and the rectangular group T. Now applying 5.3 to this subdirect product, we obtain an inclusion inverting function $\varphi: Y \rightarrow \mathcal{R}(T)$ satisfying $\bigcup_{\alpha \in Y} \alpha\varphi = T$ and for which

$$S = \{(\alpha,a) \in Y \times T \,|\, a \in \alpha\varphi\}.$$

For $A \in \mathcal{R}(T)$, let A_L, A_G and A_R be the projections of A in L, G, and R, respectively. It is clear that A is a subdirect product of A_L, A_G and A_R. Further, the regularity of A implies that of A_G which forces A_G to be a subgroup of G. Consequently 5.4 implies that $A = A_L \times A_G \times A_R$. Now defining σ, ω and τ on Y by

$$\alpha\sigma = (\alpha\varphi)_L, \qquad \alpha\omega = (\alpha\varphi)_G, \qquad \alpha\tau = (\alpha\varphi)_R$$

we quickly see that the functions σ, ω and τ satisfy all the requirements in the theorem.

For the particular cases mentioned after 5.2, we can easily specialize the last theorem to fit the given situations. The following concept will be used in the exercises below.

IV.5.6 DEFINITION. A nonempty subset A of a semigroup S is *left unitary* if $s \in S$ and $a, as \in A$ imply $s \in A$; *right unitary* is defined dually; A is *unitary* if it is both left and right unitary.

IV.5.7. Exercises

1. Show that every subdirect product of a semilattice and a periodic completely simple semigroup is a regular semigroup.

2. Let S, Y and T be as in 5.3, and on S define the relations ξ and η by:

$$(\alpha,a)\ \xi\ (\beta,b) \quad \text{if} \quad \alpha = \beta, \qquad (\alpha,a)\ \eta\ (\beta,b) \quad \text{if} \quad a = b.$$

Show that ξ and η are the least congruences on S whose quotient semigroups are a semilattice and a completely simple semigroup, respectively. Perform a similar analysis for the semigroup S in 5.5.

3. Let S be a semilattice of groups. Show that S is a sturdy semilattice of groups if and only if S does not contain a subsemigroup K of the following type: $K = A \cup B$ where A and B are disjoint semigroups, A has an idempotent, B is not a band, and for any $a \in A$, $b \in B$, we have $ab = ba = a$.

4. Show that if E_S is a left unitary subset of a regular semigroup S, then E_S is a right unitary subsemigroup of S (hence left and right unitary are equivalent on E_S for a regular semigroup S).

5. Show that if $S = (Y;S_\alpha,\psi_{\alpha,\beta})$ is a regular semigroup and E_S is unitary, then $\psi_{\alpha,\beta}$ is one-to-one on subgroups of S_α.

6. Show that if $S = [Y;S_\alpha,\psi_{\alpha,\beta}]$, where each S_α is a rectangular group and all $\psi_{\alpha,\beta}$ are one-to-one on subgroups, then E_S is unitary. Deduce that a semilattice of groups S is a subdirect product of a semilattice and a group if and only if E_S is unitary.

7. Show that in a regular semigroup S for which E_S is unitary, $ab \in E_S$ implies $ba \in E_S$.

8.* Prove that in a regular semigroup S, E_S is unitary if and only if for any $a,b \in S$ and $e \in E_S$, $aeb \in E_S$ implies $ab \in E_S$.

9. Let $S = [Y;S_\alpha,\psi_{\alpha,\beta}]$ be a regular semigroup. Show that all $\psi_{\alpha,\beta}$ are one-to-one on idempotents if and only if for any $\alpha < \beta$, $e \in E_{S_\alpha}$ implies that there exists at most one $f \in E_{S_\beta}$ such that $e < f$.

10. Prove that a semigroup S is a sturdy semilattice of rectangular groups if and only if S is completely regular, E_S is unitary, and for any $x,y \in S$ such that $N_x < N_y$, for every $f \in E_{N_y}$ there exists a unique (equivalently, at most one) $e \in E_{N_x}$ such that $e < f$, and for every $e \in E_{N_x}$ there exists at most one $f \in E_{N_y}$ such that $e < f$.

11. Find necessary and sufficient conditions on a semigroup S (in terms of semilattices of semigroups) in order that S be isomorphic to the direct product of a semilattice and a completely simple semigroup.

12.* Let $\mathcal{L}$, $\mathcal{R}$, $\mathcal{G}$ and $\mathcal{A}$ denote the classes of all left zero semigroups, right zero semigroups, groups and rectangular groups, respectively. Show that in any semigroup, the intersection of an $\mathcal{L}$-, an $\mathcal{R}$- and a $\mathcal{G}$-congruence is an $\mathcal{A}$-congruence, and conversely, every $\mathcal{A}$-congruence can be uniquely written as the intersection of an $\mathcal{L}$-, an $\mathcal{R}$- and a $\mathcal{G}$-congruence.

13.* Let $\mathcal{S}$, $\mathcal{B}$ and $\mathcal{C}$ denote the classes of all semilattices, completely simple semigroups and sturdy semilattices of completely simple semigroups, respectively. Prove that the following statements are true for any regular semigroup S. The intersection of an $\mathcal{S}$- and a $\mathcal{B}$-congruence is a $\mathcal{C}$-congruence. Conversely, every $\mathcal{C}$-congruence can be written uniquely as the intersection of an $\mathcal{S}$- and a $\mathcal{B}$-congruence.

14.* Let $\mathcal{S}$, $\mathcal{L}$, $\mathcal{R}$, and $\mathcal{G}$ have the same meaning as in the preceding two exercises, and let $\mathcal{D}$ be the class of all sturdy semilattices of rectangular groups. Prove that the following statements are true for any regular semigroup S. The intersection of an $\mathcal{S}$-, an $\mathcal{L}$-, an $\mathcal{R}$- and a $\mathcal{G}$-congruence is a $\mathcal{D}$-congruence. Conversely, every $\mathcal{D}$-congruence can be written uniquely as the intersection of an $\mathcal{S}$-, an $\mathcal{L}$-, an $\mathcal{R}$- and a $\mathcal{G}$-congruence. (*Hint*: Use the preceding two exercises.)

IV.5.8 REFERENCES:

Clifford [4], Howie and Lallement [1], Schein [6].

V

The Translational Hull

In III.1, we gave the definitions and a few simple properties of translations and the translational hull of any semigroup. In the remainder of that chapter, we used these concepts extensively in the various constructions of (ideal) extensions. The usefulness of these notions warrants a more intensive look at their properties. Hence in this chapter we first discuss some of their properties of general nature, and then consider them for several classes of semigroups. In the latter case, we are able to make some strong statements concerning the properties of the translational hull which, besides their intrinsic interest, have useful applications (in view of the results in Chapter III) to (ideal) extensions and semilattice compositions.

V.1. General Properties

We will limit ourselves here to a few elementary properties most of which have an auxiliary character. Further properties of translations can be found among exercises.

V.1.1 NOTATION. For any semigroup S, let

$$\tilde{\Lambda}(S) = \{\lambda \in \Lambda(S) \mid (\lambda,\rho) \in \Omega(S) \quad \text{for some} \quad \rho \in \mathrm{P}(S)\},$$
$$\tilde{\mathrm{P}}(S) = \{\rho \in \mathrm{P}(S) \mid (\lambda,\rho) \in \Omega(S) \quad \text{for some} \quad \lambda \in \Lambda(S)\},$$

i.e., $\tilde{\Lambda}(S)$ and $\tilde{\mathrm{P}}(S)$ are the projections of $\Omega(S)$ in $\Lambda(S)$ and $\mathrm{P}(S)$, respectively.

V.1.2 LEMMA. Left reductivity of a semigroup S is equivalent to the condition: the homomorphism $s \to \rho_s$ $(s \in S)$ is one-to-one. In such a case, the projection homomorphism $(\lambda,\rho) \to \rho$ $((\lambda,\rho) \in \Omega(S))$ is also one-to-one, and hence

$$S \cong \Delta(S) \cong \Pi(S), \qquad \Omega(S) \cong \tilde{\mathrm{P}}(S).$$

The corresponding statement is valid for right reductivity.

Proof. Exercise.

V.1.3 LEMMA. A semigroup S has a left identity if and only if every left translation of S is inner.

Proof. If e is a left identity of S, then for any $\lambda \in \Lambda(S)$ and $s \in S$, we obtain $\lambda s = \lambda(es) = (\lambda e)s = \lambda_{\lambda e}s$ which implies $\lambda = \lambda_{\lambda e}$. Conversely, if every left translation is inner, so is the identity mapping ι_S. But then $\iota_S = \lambda_e$ for some $e \in S$ and hence for any $s \in S$, we obtain $es = \lambda_e s = \iota_S s = s$.

V.1.4 COROLLARY. A semigroup S has an identity if and only if every bitranslation of S is inner.

Proof. Exercise.

V.1.5 PROPOSITION. If a semigroup S is either weakly reductive or globally idempotent, then

$$\mathcal{C}(\Omega(S)) = \{(\lambda,\rho) \in \Omega(S) \mid \lambda s = s\rho \quad \text{for all} \quad s \in S\}.$$

Proof. Let $(\lambda,\rho), (\lambda',\rho') \in \Omega(S)$ and assume that $\lambda s = s\rho$ for all $s \in S$. Using III.1.9 or III.1.10, we obtain for any $s \in S$,

$$\lambda(\lambda' s) = (\lambda' s)\rho = \lambda'(s\rho) = \lambda'(\lambda s)$$

which proves that $\lambda\lambda' = \lambda'\lambda$; one shows analogously that $\rho\rho' = \rho'\rho$ so that $(\lambda,\rho) \in \mathcal{C}(\Omega(S))$. Conversely, let $(\lambda,\rho) \in \mathcal{C}(\Omega(S))$. In particular, (λ,ρ) commutes with all inner bitranslations of S which by III.1.6 yields $\pi_{\lambda s} = \pi_{s\rho}$ for all $s \in S$. If S is weakly reductive, then $\lambda s = s\rho$ for all $s \in S$. It further follows that

$$(\lambda x)y = (x\rho)y, \qquad y(\lambda x) = y(x\rho) \qquad (x,y \in S).$$

Consequently for any $x,y \in S$, we have

$$\lambda(xy) = (\lambda x)y = (x\rho)y = x(\lambda y) = x(y\rho) = (xy)\rho,$$

which in a globally idempotent semigroup S implies $\lambda s = s\rho$ for all $s \in S$.

V.1.6 LEMMA. For a commutative semigroup S, we have $\tilde{\Lambda}(S) = \Lambda(S)$ and $\tilde{\mathrm{P}}(S) = \mathrm{P}(S)$.

Proof. Let $\lambda \in \Lambda(S)$ and define ρ on S by $s\rho = \lambda s$ for all $s \in S$. Then for any $x,y \in S$, we have

$$(xy)\rho = \lambda(xy) = \lambda(yx) = (\lambda y)x = x(\lambda y) = x(y\rho),$$
$$x(\lambda y) = x(y\rho) = (xy)\rho = (yx)\rho = y(x\rho) = (x\rho)y,$$

which shows that $(\lambda,\rho) \in \Omega(S)$ so that $\lambda \in \tilde{\Lambda}(S)$. Consequently $\tilde{\Lambda}(S) = \Lambda(S)$; one shows dually that $\tilde{\mathrm{P}}(S) = \mathrm{P}(S)$.

V.1.7 COROLLARY. If S is commutative and reductive, then $\Omega(S)$ is commutative and $\Lambda(S) \cong \mathrm{P}(S) \cong \Omega(S)$.

Proof. For any $(\lambda,\rho) \in \Omega(S)$ and $x,y \in S$, we have

$$(\lambda x)y = \lambda(xy) = \lambda(yx) = (\lambda y)x = x(\lambda y) = (x\rho)y,$$

which by reductivity implies $\lambda x = x\rho$ for all $x \in S$. Hence 1.5 implies that $\Omega(S)$ is commutative. The isomorphisms follow from 1.2 and 1.6.

V.1.8 LEMMA. Let S and T be arbitrary semigroups. The maximal subgroups of the direct product $S \times T$ are precisely the sets $G_{(e,f)} = G_e \times G_f$, where e and f are idempotents in S and T, and G_e and G_f are the corresponding maximal subgroups, respectively.

Proof. Exercise.

V.1.9 PROPOSITION. Let S be any semigroup and $(\lambda,\rho) \in \Omega(S)$ be an idempotent. By $G_{(\lambda,\rho)}$ denote the greatest subgroup of $\Omega(S)$ having

(λ,ρ) as its identity and let G_λ, G_ρ be the greatest subgroups of $\Lambda(S)$, $\mathrm{P}(S)$ having λ, ρ as identities, respectively. Then

$$G_{(\lambda,\rho)} = (G_\lambda \times G_\rho) \cap \Omega(S).$$

Proof. If G is any subgroup of $\Omega(S)$ having (λ,ρ) as its identity, then its projection G' in $\Lambda(S)$ is a subgroup of $\Lambda(S)$ contained in G_λ; similarly its projection G'' in $\mathrm{P}(S)$ is contained in G_ρ. Hence $G_{(\lambda,\rho)} \subseteq (G_\lambda \times G_\rho) \cap \Omega(S)$. Conversely, let (φ,ψ) belong to $(G_\lambda \times G_\rho) \cap \Omega(S)$. Then for any $x,y \in S$, we obtain

$$\begin{aligned} x(\varphi^{-1}y) &= x[(\lambda\varphi^{-1})y] = x[\lambda(\varphi^{-1}y)] \\ &= (x\rho)(\varphi^{-1}y) = [x(\psi^{-1}\psi)](\varphi^{-1}y) \\ &= [(x\psi^{-1})\psi](\varphi^{-1}y) = (x\psi^{-1})[\varphi(\varphi^{-1}y)] \\ &= (x\psi^{-1})[(\varphi\varphi^{-1})y] = (x\psi^{-1})(\lambda y) \\ &= [(x\psi^{-1})\rho]y = [x(\psi^{-1}\rho)]y = (x\psi^{-1})y. \end{aligned}$$

This proves that $(\varphi^{-1},\psi^{-1}) \in \Omega(S)$. But then $(G_\lambda \times G_\rho) \cap \Omega(S)$ is itself a group and therefore is the greatest subgroup of $\Omega(S)$ having (λ,ρ) as its identity.

The following concept will prove useful.

V.1.10 DEFINITION. Let A be a subsemigroup of a semigroup S. The greatest subsemigroup of S having A as an ideal is the *idealizer of A in S*, to be denoted by $i_S(A)$.

It is easy to verify that

$$i_S(A) = \{s \in S \mid sa, as \in A \quad \text{for all} \quad a \in A\}.$$

V.1.11 PROPOSITION. If S is a weakly reductive semigroup, then $\Omega(S)$ is the idealizer of $\Pi(S)$ in $\Lambda(S) \times \mathrm{P}(S)$, and $\Pi(S)$ is a densely embedded ideal of $\Omega(S)$.

Proof. For convenience let $\mathrm{T} = i_{\Lambda(S)\times \mathrm{P}(S)}(\Pi(S))$. Since $\Pi(S)$ is an ideal of $\Omega(S)$, we certainly have $\Omega(S) \subseteq \mathrm{T}$.

Conversely, let $(\lambda,\rho) \in \mathrm{T}$. Then for any $a \in S$, both $(\lambda,\rho)\pi_a$ and $\pi_a(\lambda,\rho)$ belong to $\Pi(S)$. Hence for every $a \in S$ there exists $b \in S$ such that $(\lambda,\rho)\pi_a = \pi_b$. Then for any $x,c \in S$, we have

$$[(c\rho)a]x = [(c\rho)\rho_a]x = (c\rho_b)x = c(\lambda_b x)$$
$$= c[(\lambda\lambda_a)x] = c[\lambda(\lambda_a x)]$$
$$= c[\lambda(ax)] = [c(\lambda a)]x.$$

Further, $\pi_c(\lambda,\rho) = \pi_d$ for some $d \in S$, whence

$$x[(c\rho)a] = [(xc)\rho]a = [(x\rho_c)\rho]a = (x\rho_d)a$$
$$= x(\lambda_d a) = x[(\lambda_c\lambda)a] = x[c(\lambda a)].$$

Now by the weak reductivity of S, $(c\rho)a = c(\lambda a)$. Consequently $(\lambda,\rho) \in \Omega(S)$ proving that $\mathrm{T} \subseteq \Omega(S)$.

The last statement of the proposition follows easily from III.5.9.

An analogue of the preceding proposition for left translations is provided by the following result.

V.1.12 PROPOSITION. If S is a right reductive semigroup, then $\tilde{\Lambda}(S)$ is the idealizer of $\Gamma(S)$ in $\Lambda(S)$, and $\Gamma(S)$ is a densely embedded ideal of $\tilde{\Lambda}(S)$.

Proof. Right reductivity immediately implies that the projection homomorphism of $\Omega(S)$ into $\Lambda(S)$ is one-to-one. Consequently it is an isomorphism of $\Omega(S)$ onto $\tilde{\Lambda}(S)$ mapping $\Pi(S)$ onto $\Gamma(S)$, which by 1.11 implies that $\Gamma(S)$ is a densely embedded ideal of $\tilde{\Lambda}(S)$. In addition $\tilde{\Lambda}(S) \subseteq i_{\Lambda(S)}(\Gamma(S))$. To prove the converse, let $\lambda \in \Lambda(S)$ have the property that for any $a \in S$, $\lambda_a\lambda \in \Gamma(S)$. Thus for every $a \in S$ there exists, by right reductivity, a unique $b \in S$ such that $\lambda_a\lambda = \lambda_b$. This makes it possible to define a function ρ on S by the requirement: $\lambda_a\lambda = \lambda_{a\rho}$. Thus for any $x,y \in S$, we obtain

$$\lambda_{(xy)\rho} = \lambda_{xy}\lambda = (\lambda_x\lambda_y)\lambda = \lambda_x(\lambda_y\lambda) = \lambda_x\lambda_{y\rho} = \lambda_{x(y\rho)},$$

$$\lambda_{(x\rho)y} = \lambda_{x\rho}\lambda_y = (\lambda_x\lambda)\lambda_y = \lambda_x(\lambda\lambda_y) = \lambda_x\lambda_{\lambda y} = \lambda_{x(\lambda y)},$$

which by right reductivity yields $(xy)\rho = x(y\rho)$, $(x\rho)y = x(\lambda y)$. Consequently $(\lambda,\rho) \in \Omega(S)$ showing that $\lambda \in \tilde{\Lambda}(S)$. But then $i_{\Lambda(S)}(\Gamma(S)) \subseteq \tilde{\Lambda}(S)$ as required.

We now turn to types of extension of arbitrary semigroups.

V.1.13 PROPOSITION. For any semigroup S, a subset T of $\Omega(S)$ is the type of some extension of S if and only if T is a subsemigroup of $\Omega(S)$ containing $\Pi(S)$ in which any two elements are permutable.

Proof. Necessity is a consequence of III.1.12. Conversely, let T be a subset of $\Omega(S)$ satisfying the above conditions. Let $V = \langle S,Q;\theta\rangle$ where Q is equal to T with a zero adjoined and $\theta: Q^* = \mathrm{T} \to \mathrm{T} \subseteq \Omega(S)$ is the identity mapping. Then for $\tau = \tau(V:S)$, we have $\tau^\omega = \theta^\omega = \omega$ for all $\omega \in T$ and $\tau^s = \pi_s$ for all $s \in S$, which implies that $\mathrm{T}(V:S) = \mathrm{T}$.

V.1.14 COROLLARY. If a semigroup S is either weakly reductive or globally idempotent, then the types of extension of S coincide with the subsemigroups of $\Omega(S)$ containing $\Pi(S)$.

Proof. This follows from 1.13 and III.1.9 or III.1.10.

V.1.15 PROPOSITION. Every set of permutable bitranslations of a semigroup S is contained in a maximal one.

Proof. Use Zorn's lemma; the details are left as an exercise.

V.1.16 PROPOSITION. The semigroup generated by a set of permutable bitranslations of S is a semigroup of permutable bitranslations.

Proof. Let A be a set of permutable bitranslations of S and B be the semigroup generated by A. Let $\omega,\sigma \in B$. Then $\omega = \omega_1 \ldots \omega_m$ and $\sigma = \sigma_1 \ldots \sigma_n$ where $\omega_i = (\alpha_i,\beta_i)$ and $\sigma_i = (\varphi_i,\psi_i)$, $i = 1,\ldots,m$, $j = 1,\ldots,n$. Hence $\omega = (\alpha_1 \ldots \alpha_m,\beta_1 \ldots \beta_m)$ and $\sigma = (\varphi_1 \ldots \varphi_n,\psi_1 \ldots \psi_n)$. We proceed by induction on m and n to show that $\alpha_1 \ldots \alpha_m$ and $\psi_1 \ldots \psi_n$ permute. Notice that if $n = 1$ and $m \geq 1$ this result follows easily from the hypothesis that the bitranslations in A are permutable (induction). Suppose that the left component of any element of B is permutable with the right component of every element of B which can be written as a product of less than n elements of A. Then for each $x \in S$, we have

$$\begin{aligned}(\alpha_1 \ldots \alpha_m x)(\psi_1 \ldots \psi_n) &= [(\alpha_1 \ldots \alpha_m x)(\psi_1 \ldots \psi_{n-1})]\psi_n \\ &= [(\alpha_1 \ldots \alpha_m)(x\psi_1 \ldots \psi_{n-1})]\psi_n \\ &= (\alpha_1 \ldots \alpha_n)(x\psi_1 \ldots \psi_n).\end{aligned}$$

Here we have applied the induction hypothesis twice. One shows analogously that

$$(\varphi_1 \ldots \varphi_n x)(\beta_1 \ldots \beta_m) = (\varphi_1 \ldots \varphi_n)(x\beta_1 \ldots \beta_m).$$

V.1.17 COROLLARY. Every bitranslation of a semigroup S is permutable with every inner bitranslation. Every maximal set of permutable bitranslations is a type of extension and contains $\Pi(S)$ as an ideal.

Proof. For every $(\lambda,\rho) \in \Omega(S)$ and $a,x \in S$, we obtain

$$(\lambda_a x)\rho = (ax)\rho = a(x\rho) = \lambda_a(x\rho),$$

and dually $(\lambda x)\rho_a = \lambda(x\rho_a)$. Thus $\Pi(S)$ is contained in every maximal set of permutable bitranslations.

If D is a maximal set of permutable bitranslations, then D is a semigroup by 1.16 and is a type by 1.13. Since $\Pi(S)$ is an ideal of $\Omega(S)$, it is an ideal of any type, again by 1.13.

V.1.18 COROLLARY. The union Ψ of all types of extensions of a semigroup S coincides with the set of all bitranslations $(\lambda,\rho) \in \Omega(S)$ for which λ is permutable with ρ.

Proof. If $(\lambda,\rho) \in \Psi$, then $(\lambda,\rho) \in$ T for some type T, which by 1.13 implies that λ and ρ are permutable. Conversely, if $(\lambda,\rho) \in \Omega(S)$ and λ and ρ are permutable, then $\{(\lambda,\rho)\}$ is a set of permutable bitranslations and is therefore contained in a (maximal) type by 1.17 so that $(\lambda,\rho) \in \Psi$.

V.1.19. Exercises

1. Show that the following conditions on a semigroup S are equivalent.
   i) S is a left zero semigroup.
   ii) Every transformation (written on the left) is a left translation.
   iii) The identity function (written on the right) is the only right translation.

2. Prove that if S is the free semigroup on a set, then $\Omega(S) \cong S^1$.

3. Let S be a commutative reductive semigroup. Show that $\Lambda(S)$ is the greatest commutative semigroup of transformations on S (written on the left) containing $\Gamma(S)$.

4. Show that a semigroup S is completely simple if and only if the idealizer of any cyclic subsemigroup of S is contained in some subgroup of S.

5. Let Y be a semilattice and T be any semigroup. Let $\mathcal{S}(T)$ be the set of all subsemigroups of T. Let $\varphi: Y \to \mathcal{S}(T)$ be a mapping satisfying: (i) $(\alpha\varphi)(\beta\varphi) \subseteq (\alpha\beta)\varphi$, (ii) $T = \bigcup_{\alpha \in Y} \alpha\varphi$. Let

$$S = \{(\alpha,t) \in Y \times T \mid t \in \alpha\varphi\}$$

with coordinatewise multiplication. Show that S is a subdirect product of Y and T, and conversely that every subdirect product of Y and T contained in $Y \times T$ can be so constructed.

6. Prove that every regular subsemigroup of a completely simple semigroup coincides with its idealizer.

7. Show that a nontrivial left zero semigroup is right cancellative but is not left reductive.

8. Let S be a semigroup, $(\lambda,\rho) \in \Omega(S)$ be such that λ and ρ are permutable, and let $\mathrm{T} = \Pi(S) \cup \{(\lambda,\rho)^n \mid n = 1,2, \ldots\}$. Show that T is the type of an extension of S (cf. 1.18).

9. Show that every weakly reductive globally idempotent semigroup is a subdirect product of a left reductive and a right reductive semigroup.

10. Show that a subdirect product of (left, right, weakly) reductive semigroups has the same property.

11. Show that every subsemigroup of a group G is a densely embedded ideal of its idealizer in G.

12. Let V be an inflation of a regular semigroup S. Express left translations (bitranslations) of V in terms of left translations (bitranslations) of S.

13.* Prove that the following conditions on a semigroup S are equivalent.

i) The idealizer of every cyclic subsemigroup of S is cyclic.
ii) Each subsemigroup of S coincides with its idealizer.
iii) S is a periodic completely simple semigroup.

V.1.20 REFERENCES: Goralčík [1], Goralčík and Hedrlín [1], Hedrlín and Goralčík [1], Johnson [1], Schein [2], [4], Ševrin [2], Širjaev [1], Šutov [1], Tamura [2], [3], [7], Tamura and Graham [1], Zupnik [1].

V.2. Isomorphisms and Automorphisms

We are interested here in isomorphisms between the translational hulls of two semigroups, especially in extending isomorphisms of the inner parts to the whole translational hull. As a particular case, we consider automorphisms of the translational hull.

V.2.1 THEOREM. Let θ be an isomorphism of a semigroup S onto a semigroup T. For $\lambda \in \Lambda(S)$, $\rho \in \mathrm{P}(S)$, define $\bar{\lambda}$ and $\bar{\rho}$ by:

$$\bar{\lambda}t = [\lambda(t\theta^{-1})]\theta, \qquad t\bar{\rho} = [(t\theta^{-1})\rho]\theta \qquad (t \in T).$$

Then $\lambda \to \bar{\lambda}$ and $\rho \to \bar{\rho}$ are isomorphisms of $\Lambda(S)$ onto $\Lambda(T)$ and of $\mathrm{P}(S)$ onto $\mathrm{P}(T)$, respectively. The mapping

$$\bar{\theta}:(\lambda,\rho) \to (\bar{\lambda},\bar{\rho}) \qquad ((\lambda,\rho) \in \Omega(S))$$

is an isomorphism of $\Omega(S)$ onto $\Omega(T)$ with the properties:

$$\pi_s\bar{\theta} = \pi_{s\theta}, \qquad (\omega s)\theta = (\omega\bar{\theta})(s\theta), \qquad (s\omega)\theta = (s\theta)(\omega\bar{\theta})$$
$$(s \in S, \ \omega \in \Omega(S)).$$

Proof. For any $x,y \in T$, we obtain

$$\begin{aligned}\bar{\lambda}(xy) &= \{\lambda[(xy)\theta^{-1}]\}\theta = \{\lambda[(x\theta^{-1})(y\theta^{-1})]\}\theta \\ &= \{[\lambda(x\theta^{-1})](y\theta^{-1})\}\theta = [\lambda(x\theta^{-1})]\theta y = (\bar{\lambda}x)y,\end{aligned}$$

analogously $(xy)\bar{\rho} = x(y\bar{\rho})$. Furthermore, if $(\lambda,\rho) \in \Omega(S)$, then

$$x(\bar{\lambda}y) = x[\lambda(y\theta^{-1})]\theta = \{(x\theta^{-1})[\lambda(y\theta^{-1})]\}\theta$$
$$= \{[(x\theta^{-1})\rho](y\theta^{-1})\}\theta = [(x\theta^{-1})\rho]\theta y = (x\bar{\rho})y,$$

proving that $(\bar{\lambda},\bar{\rho}) \in \Omega(T)$.

Next let $\lambda,\varphi \in \Lambda(S)$; then for any $t \in T$, we have

$$(\bar{\lambda}\bar{\varphi})t = \bar{\lambda}(\bar{\varphi}t) = \bar{\lambda}[\varphi(t\theta^{-1})]\theta = [\lambda\varphi(t\theta^{-1})]\theta = \overline{\lambda\varphi}t,$$

proving that $\bar{\lambda}\bar{\varphi} = \overline{\lambda\varphi}$. Similarly for $\rho,\psi \in \mathrm{P}(S)$, we obtain $\bar{\rho}\bar{\psi} = \overline{\rho\psi}$; in particular $\bar{\theta}$ is a homomorphism.

If $\varphi \in \Lambda(T)$ and $\psi \in \mathrm{P}(T)$, then the above calculation shows that λ and ρ defined by

$$\lambda s = [\varphi(s\theta)]\theta^{-1}, \qquad s\rho = [(s\theta)\rho]\theta^{-1} \qquad (s \in S),$$

must satisfy $\lambda \in \Lambda(S)$ and $\rho \in \mathrm{P}(S)$, and clearly $\bar{\lambda} = \varphi$ and $\bar{\rho} = \psi$, proving that all the three functions in the theorem are onto functions. A simple calculation shows that they are also one-to-one.

Furthermore, for any $s \in S$ and $t \in T$, we have

$$\overline{\lambda_s}t = [\lambda_s(t\theta^{-1})]\theta = [s(t\theta^{-1})]\theta = (s\theta)t = \lambda_{s\theta}t$$

so that $\overline{\lambda_s} = \lambda_{s\theta}$, and dually $\overline{\rho_s} = \rho_{s\theta}$, establishing the equality $\pi_s\bar{\theta} = \pi_{s\theta}$. Finally, for any $\omega = (\lambda,\rho) \in \Omega(S)$ and $s \in S$, we obtain

$$(\omega s)\theta = (\lambda s)\theta = [\lambda(s\theta\theta^{-1})]\theta = \bar{\lambda}(s\theta) = (\omega\bar{\theta})(s\theta)$$

and dually $(s\omega)\theta = (s\theta)(\omega\bar{\theta})$.

V.2.2 COROLLARY. Let T be a semigroup which is either weakly reductive or globally idempotent and let S be any semigroup. If ξ and η are isomorphisms of $\Omega(S)$ onto $\Omega(T)$ which agree on $\Pi(S)$ and map $\Pi(S)$ onto $\Pi(T)$, then $\xi = \eta$. In particular, for an isomorphism θ of S onto T, $\bar{\theta}$ is the unique isomorphism σ of $\Omega(S)$ onto $\Omega(T)$ for which $\pi_s\sigma = \pi_{s\theta}$ for all $s \in S$.

Proof. Let ξ and η be as above. For every $t \in T$, there exists $s \in S$ such that $\pi_s\xi = \pi_s\eta = \pi_t$. Hence for any $\omega \in \Omega(S)$,

$$(\omega\xi)\pi_t = (\omega\xi)(\pi_s\xi) = (\omega\pi_s)\xi = (\omega\pi_s)\eta = (\omega\eta)(\pi_s\eta) = (\omega\eta)\pi_t$$

since $\Pi(S)$ is an ideal of $\Omega(S)$, and dually one shows that $\pi_t(\omega\xi) = \pi_t(\omega\mathfrak{N})$. In view of III.1.11, we conclude $\omega\xi = \omega\eta$, and thus $\xi = \eta$. The second statement of the corollary now follows from $\pi_s\bar{\theta} = \pi_{s\theta}$ for all $s \in S$ in 2.1.

V.2.3 COROLLARY. If S and T are weakly reductive semigroups, then every isomorphism of $\Pi(S)$ onto $\Pi(T)$ can be uniquely extended to an isomorphism of $\Omega(S)$ onto $\Omega(T)$.

Proof. In light of 2.2, it suffices to observe that for a weakly reductive semigroup S, $\pi: S \to \Pi(S)$ furnishes an isomorphism of S onto $\Pi(S)$.

We denote by $\mathfrak{A}(S)$ the *group of automorphisms* of a semigroup S. Note that the *order* $o(g)$ of an element g of a group G is the order of the cyclic subgroup of G generated by g.

V.2.4 THEOREM. Let S be a semigroup. If $\theta \in \mathfrak{A}(S)$, then $o(\theta) \geq o(\bar{\theta})$; the equality holds if S is either weakly reductive or globally idempotent. In addition, if S is weakly reductive, then every automorphism of $\Pi(S)$ can be uniquely extended to an automorphism of $\Omega(S)$, and the two have the same order.

Proof. Let $\theta \in \mathfrak{A}(S)$ and $(\lambda,\rho) \in \Omega(S)$. Then a simple inductive proof shows that for any positive integer n, we have $(\lambda,\rho)\bar{\theta}^n = (\varphi_n,\psi_n)$ where

$$\varphi_n s = [\lambda(s\theta^{-n})]\theta^n, \qquad s\psi_n = [(s\theta^{-n})\rho]\theta^n \qquad (s \in S).$$

Hence if $\theta^n = \iota_S$, we obtain $(\varphi_n,\psi_n) = (\lambda,\rho)$ and thus $\bar{\theta}^n = \iota_{\Omega(S)}$, proving that $o(\theta) \geq o(\bar{\theta})$. Further, suppose $\bar{\theta}^n = \iota_{\Omega(S)}$; then

$$[\lambda(s\theta^{-n})]\theta^n = \lambda s, \qquad [(s\theta^{-n})\rho]\theta^n = s\rho \qquad (s \in S).$$

In particular, for any $x,y \in S$,

$$xy = \lambda_x y = [\lambda_x(y\theta^{-n})]\theta^n = [x(y\theta^{-n})]\theta^n = (x\theta^n)y$$

and similarly $yx = y(x\theta^n)$. Thus if S is weakly reductive, then $x\theta^n = x$ for all $x \in S$ and hence $\theta^n = \iota_S$, proving that $o(\theta) \leq o(\bar{\theta})$. From this calculation it also follows that $xy = x(y\theta^n) =$

$(x\theta^n)(y\theta^n) = (xy)\theta^n$, and hence in a globally idempotent semigroup S, $\theta^n = \iota_S$. Thus in this case also $o(\theta) \leq o(\bar{\theta})$.

The last two statements of the theorem follow from 2.3 and what we have already proved.

We can sharpen the above results for certain special classes of semigroups.

V.2.5 PROPOSITION. Let S be a weakly reductive semigroup which is either simple or 0-simple. Then $\Pi(S)$ is contained in every nonzero ideal of $\Omega(S)$, every automorphism of $\Omega(S)$ maps $\Pi(S)$ onto itself, and the mapping $\theta \rightarrow \bar{\theta}$ is an isomorphism of $\mathfrak{A}(S)$ onto $\mathfrak{A}(\Omega(S))$.

Proof. If S is simple, then $\Pi(S)$ is also since $S \cong \Pi(S)$. By I.3.11, $\Pi(S)$ is the kernel of $\Omega(S)$. Assume that S is 0-simple. Since $S \cong \Pi(S)$, we have that $\Pi(S)$ is 0-simple. Let I be an ideal of $\Omega(S)$ for which $\Pi(S) \cap I \neq \Pi(S)$. Then $\Pi(S) \cap I = \pi_0$ since $\Pi(S) \cap I$ is an ideal of $\Pi(S)$ which is 0-simple. Then for any $s \in S$ and $\omega \in I$, we have $\omega\pi_s, \pi_s\omega \in \Pi(S) \cap I = \pi_0$, and hence $\omega\pi_s = \pi_0\pi_s$, $\pi_s\omega = \pi_s\pi_0$. These equalities for all $s \in S$ imply $\omega = \pi_0$ by III.1.11. Consequently $\Pi(S)$ is contained in every nonzero ideal of $\Omega(S)$. This proves the first statement of the proposition; the second statement follows from the first since the property of "being contained in all nonzero ideals" is evidently preserved by every automorphism. Combining this with the last assertion of 2.4, we see that the function which to each automorphism of $\Omega(S)$ associates its restriction to $\Pi(S)$ is an isomorphism of $\mathfrak{A}(\Omega(S))$ onto $\mathfrak{A}(\Pi(S))$. But then $\theta \rightarrow \bar{\theta}$ must be an isomorphism of $\mathfrak{A}(S)$ onto $\mathfrak{A}(\Omega(S))$.

We will now briefly discuss inner automorphisms.

V.2.6 DEFINITION. If S is a semigroup with identity and a is an element of the group of units of S, the function ϵ_a defined by:

$$s\epsilon_a = a^{-1}sa \qquad (s \in S)$$

is the *inner automorphism* of S induced by a.

We cannot speak of inner automorphisms for a semigroup without identity, but we can find a substitute as follows. We denote by $\Sigma(S)$ the *group of units of* $\Omega(S)$ for any semigroup S.

V.2.7 LEMMA. Let S be a semigroup, let $(\lambda,\rho) \in \Sigma(S)$ and assume that λ and ρ are permutable. The function $\delta_{(\lambda,\rho)}$ defined by

$$s\delta_{(\lambda,\rho)} = (\lambda^{-1}s)\rho \qquad (s \in S)$$

is an automorphism of S.

Proof. For a given $\delta_{(\lambda,\rho)}$, both λ and ρ are permutations of S and hence so is $\delta_{(\lambda,\rho)}$. For any $x,y \in S$, we obtain

$$(\lambda^{-1}x)y = \lambda^{-1}\{\lambda[(\lambda^{-1}x)y]\} = \lambda^{-1}[(\lambda\lambda^{-1}x)y] = \lambda^{-1}(xy)$$

and $\lambda^{-1} \in \Lambda(S)$, hence

$$\begin{aligned}(xy)\delta_{(\lambda,\rho)} &= [\lambda^{-1}(xy)]\rho = [(\lambda^{-1}x)y]\rho = (\lambda^{-1}x)(y\rho)\\ &= (\lambda^{-1}x)[(\lambda\lambda^{-1}y)\rho] = (\lambda^{-1}x)\{\lambda[(\lambda^{-1}y)\rho]\}\\ &= [(\lambda^{-1}x)\rho][(\lambda^{-1}y)\rho] = (x\delta_{(\lambda,\rho)})(y\delta_{(\lambda,\rho)}),\end{aligned}$$

proving that $\delta_{(\lambda,\rho)}$ is also a homomorphism.

V.2.8 DEFINITION. The mapping $\delta_{(\lambda,\rho)}$ defined above is the *generalized inner automorphism* of S induced by (λ,ρ).

By $\mathcal{G}(S)$ we denote the *set of all generalized inner automorphisms of* S. We have seen in the proof of 2.7 that for $\delta_{(\lambda,\rho)}$, λ^{-1} is a left translation. Further, for any $s \in S$,

$$(\lambda^{-1}s)\rho = \lambda^{-1}\lambda[(\lambda^{-1}s)\rho] = \lambda^{-1}[(\lambda\lambda^{-1}s)\rho] = \lambda^{-1}(s\rho),$$

and hence we can write $s\delta_{(\lambda,\rho)} = \lambda^{-1}s\rho$ if we wish. It is clear that in the case that S has an identity, generalized inner automorphisms coincide with inner automorphisms.

A part of the following result is an analogue of a familiar theorem in group theory.

V.2.9 THEOREM. Let S be a semigroup which is either weakly reductive or globally idempotent. Then $\mathcal{G}(S)$ is a subgroup of $\mathcal{A}(S)$ and the mapping χ defined by

$$\chi\colon(\lambda,\rho) \to \delta_{(\lambda,\rho)} \qquad ((\lambda,\rho) \in \Sigma(S))$$

is a homomorphism of $\Sigma(S)$ onto $\mathcal{G}(S)$ with kernel $\Sigma(S) \cap \mathcal{C}(\Omega(S))$ so that

$$\Sigma(S)/[\Sigma(S) \cap \mathcal{C}(\Omega(S))] \cong \mathcal{G}(S).$$

In addition $\overline{\delta_{(\lambda,\rho)}} = \epsilon_{(\lambda,\rho)}$.

Proof. If $(\lambda,\rho) \in \Sigma(S)$, then by III.1.9 or III.1.10, λ and ρ are permutable and thus $\delta_{(\lambda,\rho)}$ is defined. Let $(\lambda,\rho),(\varphi,\psi) \in \Sigma(S)$. Then $(\lambda^{-1},\rho^{-1}) = (\lambda,\rho)^{-1} \in \Sigma(S)$ which implies that λ^{-1} is permutable with ψ. For any $s \in S$, we obtain

$$\begin{aligned} s\delta_{(\lambda,\rho)}\delta_{(\varphi,\psi)} &= \varphi^{-1}\{[(\lambda^{-1}s)\rho]\psi\} = \varphi^{-1}\{[\lambda^{-1}(s\rho)]\psi\} \\ &= \varphi^{-1}[\lambda^{-1}(s\rho\psi)] = (\varphi^{-1}\lambda^{-1})(s\rho\psi) \\ &= (\lambda\varphi)^{-1}s(\rho\psi) \end{aligned}$$

which implies $\delta_{(\lambda,\rho)}\,\delta_{(\varphi,\psi)} = \delta_{(\lambda\varphi,\rho\psi)} = \delta_{(\lambda,\rho)(\varphi,\psi)}$ and proves that χ is a homomorphism and $\mathcal{G}(S)$ is a group. Further, for any $(\lambda,\rho) \in \Sigma(S)$, $\delta_{(\lambda,\rho)} = (\iota_{\Omega(S)}, \iota_{\Omega(S)})$ is equivalent to $\lambda^{-1}s\rho = s$ for all $s \in S$, which can be written as $\lambda s = s\rho$ for all $s \in S$. By 1.5, the latter is equivalent to $(\lambda,\rho) \in \Sigma(S) \cap \mathcal{C}(\Omega(S))$. This proves the assertion concerning the kernel; the isomorphism is a consequence of the fundamental homomorphism theorem for groups.

Let $(\lambda,\rho) \in \Sigma(S)$ and $(\varphi,\psi) \in \Omega(S)$. Then $(\varphi,\psi)\overline{\delta_{(\lambda,\rho)}} = (\varphi',\psi')$, where for all $s \in S$,

$$\varphi's = [\varphi(s\overset{-1}{\delta}_{(\lambda,\rho)})]\delta_{(\lambda,\rho)} = \lambda^{-1}[\varphi(\lambda s\rho^{-1})]\rho = (\lambda^{-1}\varphi\lambda)s$$

and analogously $s\psi' = s(\rho^{-1}\psi\rho)$. Consequently $(\varphi',\psi') = (\lambda,\rho)^{-1}(\varphi,\psi)(\lambda,\rho) = (\varphi,\psi)\epsilon_{(\lambda,\rho)}$ as required.

We always have $\Sigma(S) \cap \mathcal{C}(\Omega(S)) \subseteq \mathcal{C}(\Sigma(S))$; in 5.7 we will encounter an example showing that the inclusion may be strict.

V.2.10. Exercises

1. Show that for any semigroup S, the mapping $\sigma:\theta \to \bar{\theta}$ $(\theta \in \mathcal{A}(S))$ is a homomorphism of $\mathcal{A}(S)$ into $\mathcal{A}(\Omega(S))$. Also show that if S is

weakly reductive or globally idempotent, then $\sigma|_{\mathcal{G}(S)}$ is an isomorphism of $\mathcal{G}(S)$ onto $\mathcal{A}(\Omega(S))$.

2. Let S be any semigroup for which any two elements of $\Sigma(S)$ are permutable. A nonempty subset K of S is called a *conjugate* of a subset K' of S if there exists $(\lambda,\rho) \in \Sigma(S)$ such that $\lambda^{-1}K\rho = K'$. Show that the set

$$\{(\lambda,\rho) \in \Sigma(S) \mid \lambda^{-1}K\rho = K\}$$

is a subgroup of $\Sigma(S)$ whose index in $\Sigma(S)$ equals the number of conjugates of K.

3. Let S and S' be 0-simple or simple weakly reductive semigroups and φ be an isomorphism of T onto T', where $\Pi(S) \subseteq T \subseteq \Omega(S)$ and $\Pi(S') \subseteq T' \subseteq \Omega(S')$. Show that there exists a unique isomorphism θ of S onto S' such that $\bar{\theta}\,|_T = \varphi$.

V.2.11 REFERENCES: Croisot [2], Dubreil [1], Gluskin [3], Thierrin [5].

V.3. Rees Matrix Semigroups

We have defined in IV.2.6 a Rees matrix semigroup over a group. Now we wish to find a suitable isomorphic copy of its translational hull. These semigroups are closely related to a more general class of semigroups with zero. Since the latter class of semigroups plays a very important role in view of the Rees theorem (which we have not discussed in this book) and since it takes only a little more effort to find its translational hull, we will find an isomorphic copy for the semigroups of this wider class. Note that "a group with a zero adjoined" is usually abbreviated to "a group with zero" and denoted by G^0.

V.3.1 DEFINITION. Let G^0 be a group with zero, let I and M be nonempty sets, and $P : \mathrm{M} \times I \to G^0$ be any function satisfying the following requirement: for each $\mu \in \mathrm{M}$, there exists $i \in I$ such that $p_{\mu i} \neq 0$, and for each $j \in I$, there exists $\nu \in \mathrm{M}$ such that $p_{\nu j} \neq 0$. Let T be the set $I \times G^0 \times \mathrm{M}$ together with the multiplication

$$(i,a,\mu)(j,b,\nu) = (i,ap_{\mu j}b,\nu). \tag{1}$$

Then T is a semigroup and the set $J = \{(i,0,\mu) \mid i \in I,\ \mu \in M\}$ is an ideal of T. The semigroup $S = T/J$ is called the *Rees* $I \times M$ *matrix semigroup over the group with zero* G^0 *with the sandwich matrix* P and is denoted by $S = \mathfrak{M}^0(I,G,M;P)$.

We will usually call such a semigroup a *Rees matrix semigroup with zero*, or simply a *Rees matrix semigroup*, if there is no danger of confusion (the semigroup defined in IV.2.6 is sometimes called a *Rees matrix semigroup without zero*). Forming the Rees quotient T/J is usually referred to as identifying all triples with second entry zero. Strictly speaking, multiplication in S is defined by

$$(i,a,\mu)(j,b,\nu) = \begin{cases} (i,ap_{\mu j}b,\nu) & \text{if } p_{\mu j} \neq 0, \\ 0 \text{ otherwise,} \end{cases}$$

$$(i,a,\mu)0 = 0(i,a,\mu) = 00 = 0.$$

Hence, to simplify notation, we will write all elements of S in the form (i,a,μ), where the zero is written as $(i,0,\mu)$ with arbitrary i, μ (or as 0), and adopt multiplication in (1). As usual, the semigroup S can be considered as defined on the set $(I \times G \times M) \cup 0$.

These semigroups also admit a matrix interpretation. We modify the discussion after IV.2.6 only by admitting the $I \times M$ matrix all of whose entries are zero. Thus our semigroup consists of all $I \times M$ matrices over the group with zero G^0 having at most one nonzero entry and multiplying by the rule $A * B = APB$.

If $S = \mathfrak{M}^0(I,G,M;P)$ and 0 is a completely prime ideal of S, then S^* is a semigroup, in fact, a Rees matrix semigroup without zero. This is equivalent to the sandwich matrix P having no zero entries. Conversely, if $S = \mathfrak{M}(I,G,M;P)$, then $\mathfrak{M}^0(I,G,M;P)$ can be interpreted as S with a zero adjoined and 0 is a completely prime ideal of this semigroup. Therefore, in studying the properties of both kinds of semigroups, it suffices in most cases to prove results about a Rees matrix semigroup with zero and then obtain the corresponding results for a Rees matrix semigroup without zero. However, it is sometimes easier to consider the case without zero, since the case with zero sometimes presents difficulties not present in the other case. For example, IV.3.3 has an analogue for the case with zero which is much more complicated. We will usually state a property for the case with zero and leave as an implicit (though imperative) exercise the statement and derivation of the corresponding property for the case without zero.

V.3.2 DEFINITION. Let X be a set. A function α mapping a subset Y of X into X is a *partial transformation* on X; Y is the *domain* of α, denoted by $\mathbf{d}\alpha$, and $\{x \in X \mid \alpha y = x \text{ for some } y \in Y\}$ is the *range* of α, denoted by $\mathbf{r}\alpha$; the cardinal number $|\mathbf{r}\alpha|$ is the *rank* of α, denoted by **rank** α. For convenience, the *empty transformation*, denoted by $\varnothing$, is the mapping with $\mathbf{d}\varnothing = \mathbf{r}\varnothing = \varnothing$, the empty set.

Let $\mathfrak{F}(X)$ be the set of all partial transformations on X together with the multiplication $(\alpha\beta)x = \alpha(\beta x)$ for all $x \in \mathbf{d}\beta$ for which $\beta x \in \mathbf{d}\alpha$; otherwise $(\alpha\beta)x$ is not defined.

It is easy to verify that $\mathfrak{F}(X)$ with this multiplication is a semigroup with zero $\varnothing$ and identity ι_X; $\mathfrak{F}(X)$ is the *semigroup of all partial transformations on X written on the left.*

We can essentially repeat these definitions by writing the functions on the right. The corresponding semigroup is denoted by $\mathfrak{F}'(X)$. For example, multiplication in $\mathfrak{F}'(X)$ is given by: $x(\alpha\beta) = (x\alpha)\beta$ for all $x \in \mathbf{d}\alpha$ for which $x\alpha \in \mathbf{d}\beta$.

The semigroup $\Lambda(S)$, for a Rees matrix semigroup S, can be constructed by a device analogous to the wreath product of groups. We proceed to describe this construction.

Let X be a nonempty set and G be a group. We have denoted by $\varnothing$ the empty transformation on X; we now denote by the same symbol $\varnothing$ the "empty function" from X to G (again, $\mathbf{d}\varnothing = \mathbf{r}\varnothing = \varnothing$), only for the convenience of notation. Let φ and φ' be functions from subsets of X (including possibly the empty set) into G, written on the left, and define the product of these two functions "pointwise" by

$$(\varphi \cdot \varphi')x = (\varphi x)(\varphi' x)$$

for all $x \in X$ for which $(\varphi x)(\varphi' x)$ is defined, i.e., for all $x \in \mathbf{d}\varphi \cap \mathbf{d}\varphi'$. For $\alpha \in \mathfrak{F}(X)$ and φ as before, define φ^α as the function

$$\varphi^\alpha x = \varphi(\alpha x)$$

for all $x \in X$ for which $\varphi(\alpha x)$ is defined, i.e., for all $x \in \mathbf{d}\alpha$ such that $\alpha x \in \mathbf{d}\varphi$.

V.3.3 DEFINITION. Let P be a subsemigroup of $\mathfrak{F}(X)$ and G be a group. The *left wreath product* of P and G, denoted by $P\ w\ell\ G$, is the set

$$\{(\alpha,\varphi) \mid \alpha \in P,\ \varphi{:}\mathbf{d}\alpha \to G\}$$

together with the multiplication

$$(\alpha,\varphi)(\alpha',\varphi') = (\alpha\alpha',\varphi^{\alpha'}\cdot\varphi').$$

It is easy to see that $P\ w\ell\ G$ is closed under this multiplication (i.e., $\alpha\alpha'$ and $\varphi^{\alpha'}\cdot\varphi'$ have the same domain). That this multiplication is associative will follow from the next theorem.

Let $S = \mathfrak{M}^0(I,G,M;P)$, $\lambda \in \Lambda(S)$ and suppose that $\lambda(i,a,\mu) \neq 0$. Then there exists $j \in I$ such that $p_{\mu j} \neq 0$ and thus

$$\lambda(i,a,\mu) = \lambda[(i,a,\mu)(j,p_{\mu j}^{-1},\mu)] = [\lambda(i,a,\mu)](i,p_{\mu j}^{-1},\mu) \neq 0.$$

Hence $\lambda(i,a,\mu) = (k,b,\mu)$ for some $k \in I$, $b \in G$. In other words a left translation does not change the index in M.

As usual the identity of G will be denoted by 1. Suppose next that $\lambda(i,1,\mu) = (j,b,\mu) \neq 0$ and that $\lambda(i,1,\nu) = (k,c,\nu) \neq 0$, Then for some $m \in I$, $p_{\mu m} \neq 0$ and thus

$$\begin{aligned}(j,b,\nu) &= (j,b,\mu)(m,p_{\mu m}^{-1},\nu) = [\lambda(i,1,\mu)](m,p_{\mu m}^{-1},\nu)\\ &= \lambda[(i,1,\mu)(m,p_{\mu m}^{-1},\nu)] = \lambda(i,1,\nu) = (k,c,\nu) \neq 0.\end{aligned}$$

Consequently $j = k$ and $b = c$ which shows that the value of the first two indices of $\lambda(i,1,\mu)$ depends only on i.

In view of this discussion, we are able to define the function $\mathfrak{a}$ below.

V.3.4 THEOREM. Let $S = \mathfrak{M}^0(I,G,M;P)$. Then the function $\mathfrak{a}$ defined by

$$\mathfrak{a}\lambda = (\alpha,\varphi) \qquad (\lambda \in \Lambda(S)),$$

where

$$\begin{aligned}\mathbf{d}\alpha = \mathbf{d}\varphi &= \{i \in I \mid \lambda(i,1,\mu) \neq 0\}, \qquad (1)\\ \lambda(i,1,\mu) &= (\alpha i,\varphi i,\mu) \quad \text{if} \quad i \in \mathbf{d}\alpha,\end{aligned}$$

is an isomorphism of $\Lambda(S)$ onto $\mathfrak{F}(I)\ w\ell\ G$.

Proof. We observe that for $\mathfrak{a}\lambda = (\alpha,\varphi)$, $i \in \mathbf{d}\alpha$ and any $(i,a,\mu) \in S$, there exists $j \in I$ such that $p_{\mu j} \neq 0$, and thus

$$\begin{aligned}\lambda(i,a,\mu) &= \lambda[(i,1,\mu)(j,p_{\mu j}^{-1}a,\mu)] = [\lambda(i,1,\mu)](j,p_{\mu j}^{-1}a,\mu)\\ &= (\alpha i,\varphi i,\mu)(j,p_{\mu j}^{-1}a,\mu) = (\alpha i,(\varphi i)a,\mu).\end{aligned}$$

Consequently

$$\lambda(i,a,\mu) = \begin{cases}(\alpha i,(\varphi i)a,\mu) & \text{if} \quad i \in \mathbf{d}\alpha, \\ 0 & \text{if} \quad i \notin \mathbf{d}\alpha.\end{cases} \tag{2}$$

To show that $\mathfrak{a}$ is a homomorphism, let $\lambda,\lambda' \in \Lambda(S)$ with $\mathfrak{a}\lambda = (\alpha,\varphi)$ and $\mathfrak{a}\lambda' = (\alpha',\varphi')$. If $\alpha\alpha' \neq \varnothing$, let $i \in \mathbf{d}(\alpha\alpha')$. Then $i \in \mathbf{d}\alpha'$, $\alpha' i \in \mathbf{d}\alpha$, and using the existence of $j \in I$ such that $p_{\mu j} \neq 0$, we obtain by (2),

$$\begin{aligned}\lambda\lambda'(i,1,\mu) &= \lambda(\alpha' i,\varphi' i,\mu) = (\alpha\alpha' i,(\varphi\alpha' i)(\varphi' i),\mu) \\ &= (\alpha\alpha' i,(\varphi^{\alpha'}\cdot\varphi')i,\mu).\end{aligned}$$

If $\alpha\alpha' = \varnothing$, then $\lambda\lambda' = 0$, since otherwise there would exist $i \in I$ such that $\lambda\lambda'(i,1,\mu) \neq 0$ and the above calculation would imply that $\alpha\alpha' \neq \varnothing$. Therefore $\mathfrak{a}$ is a homomorphism. It follows from (2) that $\mathfrak{a}\lambda = \mathfrak{a}\lambda'$ implies $\lambda = \lambda'$ so that $\mathfrak{a}$ is one-to-one.

To show that $\mathfrak{a}$ is onto, let $(\alpha,\varphi) \in \mathfrak{F}(I)\ w\ell\ G$ and define λ by (2). If $i \in \mathbf{d}\alpha$ and $p_{\mu j} \neq 0$, then

$$\begin{aligned}[\lambda(i,a,\mu)](j,b,\nu) &= (\alpha i,(\varphi i)a,\mu)(j,b,\nu) = (\alpha i,(\varphi i)ap_{\mu j}b,\nu) \\ &= \lambda(i,ap_{\mu j}b,\nu) = \lambda[(i,a,\mu)(j,b,\nu)],\end{aligned}$$

while otherwise both $[\lambda(i,a,\mu)](j,b,\nu)$ and $\lambda[(i,a,\mu)(j,b,\nu)]$ are zero. We conclude that $\lambda \in \Lambda(S)$ and by (2) that $\mathfrak{a}\lambda = (\alpha,\varphi)$ which proves that $\mathfrak{a}$ is also onto. Thus $\mathfrak{a}$ is an isomorphism of $\Lambda(S)$ onto $\mathfrak{F}(I)\ w\ell\ G$.

We next construct the "right" version of the wreath product. Let X be a nonempty set and G a group; the identity of G will be denoted by 1. The empty function from X into G has the same meaning as above. For the functions ψ and ψ', written on the right, from subsets of X into G, define their product by

$$x(\psi\cdot\psi') = (x\psi)(x\psi')$$

for all $x \in \mathbf{d}\psi \cap \mathbf{d}\psi'$. For $\beta \in \mathfrak{F}'(X)$ and ψ as before, define ${}^{\beta}\psi$ as the function

$$x^{\beta}\psi = (x\beta)\psi$$

for all $x \in \mathbf{d}\beta$ such that $x\beta \in \mathbf{d}\psi$.

V.3.5 DEFINITION. Let G be a group and Q a subsemigroup of $\mathfrak{F}'(X)$. Then the *right wreath product* of G and Q, denoted by G *wr* Q, is the set

$$\{(\psi,\beta) \mid \psi{:}\mathbf{d}\beta \to G,\ \beta \in Q\}$$

together with the multiplication

$$(\psi,\beta)(\psi',\beta') = (\psi \cdot {}^{\beta}\psi', \beta\beta').$$

A discussion entirely analogous to that preceding 3.4 assures that we may define the function $\mathfrak{b}$ below.

V.3.6 THEOREM. Let $S = \mathfrak{M}^0(I,G,M;P)$. Then the function $\mathfrak{a}$ defined by

$$\rho\mathfrak{b} = (\psi,\beta) \qquad (\rho \in P(S))$$

where

$$\begin{aligned} \mathbf{d}\beta = \mathbf{d}\psi &= \{\mu \in M \mid (i,1,\mu)\rho \neq 0\}, \\ (i,1,\mu)\rho &= (i,\mu\psi,\mu\beta) \quad \text{if} \quad \mu \in \mathbf{d}\beta, \end{aligned} \tag{1}$$

is an isomorphism of $P(S)$ onto G *wr* $\mathfrak{F}'(M)$.

Proof. The proof is analogous to that of 3.4 and is left as an exercise. We only give the analogue of (1) in 3.4, viz.,

$$(i,a,\mu)\rho = \begin{cases} (i,a(\mu\psi),\mu\beta) & \text{if} \quad \mu \in \mathbf{d}\beta, \\ 0 & \text{otherwise,} \end{cases} \tag{2}$$

and ρ has the property that $\rho\mathfrak{b} = (\psi,\beta)$.

The linking of a left and a right translation can be expressed as follows.

V.3.7 PROPOSITION. Let $S = \mathfrak{M}^0(I,G,M;P)$ and let $\mathfrak{a}\lambda = (\alpha,\varphi)$ and $\rho\mathfrak{b} = (\psi,\beta)$. Then $(\lambda,\rho) \in \Omega(S)$ if and only if the following conditions hold: For any $i \in I$, $\mu \in M$,

i) $i \in \mathbf{d}\alpha,\ p_{\mu(\alpha i)} \neq 0$ if and only if $\mu \in \mathbf{d}\beta,\ p_{(\mu\beta)i} \neq 0$,

ii) $p_{\mu(\alpha i)}(\varphi i) = (\mu\psi)p_{(\mu\beta)i}$ if $i \in \mathbf{d}\alpha,\ p_{\mu(\alpha i)} \neq 0$.

Further, for $(\lambda,\rho) \in \Omega(S)$, we have $\lambda = 0$ if and only if $\rho = 0$.

Proof. The first statement follows directly from the following calculation:

$$(j,a,\mu)[\lambda(i,b,\nu)] = \begin{cases}(j,a,\mu)(\alpha i,(\varphi i)b,\nu) & \text{if } i \in \mathbf{d}\alpha \\ 0 & \text{otherwise}\end{cases}$$

$$= \begin{cases}(j,ap_{\mu(\alpha i)}(\varphi i)b,\nu) & \text{if } i \in \mathbf{d}\alpha,\ p_{\mu(\alpha i)} \neq 0, \\ 0 & \text{otherwise,}\end{cases}$$

and similarly

$$[(j,a,\mu)\rho](i,b,\nu) = \begin{cases}(j,a(\mu\psi)p_{(\mu\beta)i}b,\nu) & \text{if } \mu \in \mathbf{d}\beta,\ p_{(\mu\beta)i} \neq 0, \\ 0 & \text{otherwise.}\end{cases}$$

The proof of the second statement is left as an exercise.

Writing the elements of $\Omega(S)$ as (λ,ρ) in our present case amounts to writing $(\mathfrak{a}^{-1}(\alpha,\varphi), (\psi,\beta)\mathfrak{b}^{-1})$. To simplify the notation, it is convenient to identify (α,φ) with λ and (ψ,β) with ρ, respectively, and write the elements of $\Omega(S)$ as $((\alpha,\varphi), (\psi,\beta))$.

Condition i) in 3.7 usually presents great difficulties. However, this condition is vacuous if all entries of P are nonzero (i.e., in the case in which 0 is a completely prime ideal). Condition ii) can be written in the form

$$\mu\psi = p_{\mu(\alpha i)}(\varphi i)p^{-1}_{(\mu\beta)i} \quad \text{if} \quad i \in \mathbf{d}\alpha, p_{\mu(\alpha i)} \neq 0.$$

Thus ψ is completely determined by α,φ,β. For (α,φ) and (ψ,β) as in 3.3 and 3.5, for the sake of convenience, we let

$$\mathbf{rank}(\alpha,\varphi) = \mathbf{rank}\ \alpha, \qquad \mathbf{rank}(\psi,\beta) = \mathbf{rank}\ \beta.$$

We will now locate $\Pi(S)$ in $\Omega(S)$.

V.3.8 THEOREM. Let $S = \mathfrak{M}^0(I,G,M;P)$. For $0 \neq (\lambda,\rho) \in \Omega(S)$, we have $(\lambda,\rho) \in \Pi(S)$ if and only if rank $\mathfrak{a}\lambda$ = rank $\rho\mathfrak{b}$ = 1.

Proof. Let $\mathfrak{a}\lambda = (\alpha,\varphi)$, $\rho\mathfrak{b} = (\beta,\psi)$ and $0 \neq (\lambda,\rho) \in \Omega(S)$.

Necessity. Let $(\lambda,\rho) \in \Pi(S)$. Then by the definition of $\Pi(S)$, we have

$$\lambda = \lambda_{(j,b,\nu)}, \qquad \rho = \rho_{(j,b,\nu)}, \tag{1}$$

for some nonzero (j,b,ν) in S. Using (1) of 3.4 and (1) of 3.6, we obtain

$$\mathbf{d}\alpha = \{i \in I \mid p_{\nu i} \neq 0\}, \tag{2}$$

$$\mathbf{d}\beta = \{\mu \in \mathrm{M} : p_{\mu j} \neq 0\}. \tag{3}$$

For $i \in \mathbf{d}\alpha$, using (2) of 3.4, we have

$$\lambda(i,a,\mu) = (j,b,\nu)(i,a,\mu) = (j,bp_{\nu i}a,\mu) = (\alpha i,(\varphi i)a,\mu),$$

whence

$$\varphi i = bp_{\nu i}, \qquad \alpha i = j \qquad (i \in \mathbf{d}\alpha); \tag{4}$$

the last equality implies that **rank** $\alpha = 1$. Analogous computation shows, using (2) of 3.6, that

$$\mu\psi = p_{\mu j}b, \qquad \mu\beta = \nu \qquad (\mu \in \mathbf{d}\beta); \tag{5}$$

again the last equality implies that **rank** $\beta = 1$.

Sufficiency. By hypothesis **rank** $\alpha =$ **rank** $\beta = 1$ so that $\alpha i = j$ for all $i \in \mathbf{d}\alpha$ and some $j \in I$; similarly $\mu\beta = \nu$ for all $\mu \in \mathbf{d}\beta$ and some $\nu \in \mathrm{M}$.

Let $i \in I$ be arbitrary. If $P_{\nu i} \neq 0$, then for any $\mu \in \mathbf{d}\beta$, we have $p_{(\mu\beta)i} = p_{\nu i} \neq 0$ and hence $i \in \mathbf{d}\alpha$ by 3.7. Conversely, assume that $i \in \mathbf{d}\alpha$. Then $\alpha i = j$ and there exists $\mu \in \mathrm{M}$ such that $p_{\mu j} \neq 0$. But then $i \in \mathbf{d}\alpha$ and $p_{\mu(\alpha i)} \neq 0$ imply $p_{(\mu\beta)i} \neq 0$ by 3.7 so that $p_{\nu i} \neq 0$. This proves (2); (3) is established analogously.

Let $i \in \mathbf{d}\alpha$, $\mu \in \mathbf{d}\beta$; then by (2) and (3), we have $p_{\nu i} \neq 0$ and $p_{\mu j} \neq 0$, and by 3.7 ii), we also have $p_{\mu j}(\varphi i) = (\mu\psi)p_{\nu i}$, whence $(\varphi i)p_{\nu i}^{-1} = p_{\mu j}^{-1}(\mu\psi)$. Since j and ν are fixed, in the last equation the left hand side depends only on i, while the right hand side depends only on μ. Since $i \in \mathbf{d}\alpha$ and $\mu \in \mathbf{d}\beta$ are arbitrary, both sides must be equal to a constant, say b. Hence (4) and (5) are satisfied. Since also (2) and (3) have been established, we may reverse the steps of the first part of the proof to see that (1) also holds. Thus $(\lambda,\rho) \in \Pi(S)$.

V.3.9 COROLLARY. For $S = \mathfrak{M}^0(I,G,\mathrm{M};P)$, we have

$$\begin{aligned}\Pi(S) &= \{(\lambda,\rho) \in \Omega(S) \mid \mathbf{rank}\ \mathfrak{a}\lambda \leq 1,\ \mathbf{rank}\ \rho\mathfrak{b} \leq 1\} \\ &= \{(\lambda,\rho) \in \Omega(S) \mid \mathbf{rank}\ \mathfrak{a}\lambda = 1 = \mathbf{rank}\ \rho\mathfrak{b}\} \cup 0.\end{aligned}$$

We consider next the translational hull of a Rees matrix semigroup over a group. Most of the results will follow from the case already considered via the following lemma. The functions $\alpha,\varphi,\mathfrak{a}$, etc., have the same meaning as before even though S has no zero.

V.3.10 LEMMA. Let T be any semigroup and $S = T \cup 0$, where $0 \notin T$ and 0 acts as the zero of S. If $\lambda \in \Lambda(T)$, then the function λ' defined by

$$\lambda' x = \begin{cases} \lambda x & \text{if } \ x \in T \\ 0 & \text{if } \ x = 0 \end{cases}$$

satisfies $\lambda' \in \Lambda(S)$. Conversely, if $\xi \in \Lambda(S)$ maps T into T, then $\xi|_T \in \Lambda(T)$. The corresponding statements are valid for right translations ρ, ρ'. Further $(\lambda,\rho) \in \Omega(T)$ implies $(\lambda',\rho') \in \Omega(S)$; conversely, if $(\xi,\eta) \in \Omega(S)$ and both ξ,η map T into T, then $(\xi|_T,\eta|_T) \in \Omega(T)$. Finally

$$\Omega(T) = \{(\xi|_T,\eta|_T) \mid (\xi,\eta) \in \Omega(S),\ \xi,\eta{:}T \to T\}.$$

Proof. Exercise.

Recall the notation $\mathfrak{I}(X)$ and $\mathfrak{I}'(X)$ for the semigroups of all transformations on a set X from I.6.2.

V.3.11 THEOREM. For $S = \mathfrak{M}(I,G,\mathrm{M};P)$, the following statements are valid.

i) $\Lambda(S) \cong \mathfrak{I}(I)\ w\ell\ G$.
ii) $\mathrm{P}(S) \cong G\ wr\ \mathfrak{I}'(\mathrm{M})$.
iii) For $\mathfrak{a}\lambda = (\alpha,\varphi)$ and $\rho\mathfrak{b} = (\psi,\beta)$, we have $(\lambda,\rho) \in \Omega(S)$ if and only if $p_{\mu(\alpha i)}(\varphi i) = (\mu\psi)p_{(\mu\beta)i}$ for all $i \in I$, $\mu \in \mathrm{M}$.
iv) $\Pi(S) = \{(\lambda,\rho) \in \Omega(S) \mid \textbf{rank}\ \mathfrak{a}\lambda = \textbf{rank}\ \rho\mathfrak{b} = 1\}$.

Proof. This follows from 3.4, 3.6 and 3.9 in conjunction with 3.10.

V.3.12 COROLLARY. Let L, G and R be a left zero semigroup, a group and a right zero semigroup, respectively. The following statements hold.

i) $\Omega(L \times G \times R) \cong \mathfrak{I}(L) \times G \times \mathfrak{I}'(R)$.
ii) $\Omega(L \times G) \cong \mathfrak{I}(L) \times G$, $\Omega(G \times R) \cong G \times \mathfrak{I}'(R)$.
iii) $\Omega(L \times R) \cong \mathfrak{I}(L) \times \mathfrak{I}'(R)$, $\Omega(L) \cong \mathfrak{I}(L)$, $\Omega(R) \cong \mathfrak{I}'(R)$.

Proof. It suffices to prove item *i*); the remaining items are its obvious consequences. Hence let $S = L \times G \times R$. It is clear that

$S \cong \mathfrak{M}(I,G,\mathrm{M};P)$ where $|L| = |I|$, $|R| = |\mathrm{M}|$ and all entries of P are equal to the identity of G. Hence the condition in 3.11 iii) reduces to $\varphi i = \mu\psi$ for all $i \in I$, $\mu \in \mathrm{M}$ where $(\lambda,\rho) \in \Omega(\mathfrak{M}(I,G,\mathrm{M};P))$ and $\mathfrak{a}\lambda = (\alpha,\varphi)$, $\rho\mathfrak{b} = (\psi,\beta)$. Letting $\varphi i = \mu\psi = g \in G$, it is easy to verify that the function χ defined on $\Omega(\mathfrak{M}(I,G,\mathrm{M};P))$ by $\chi\colon (\lambda,\rho) \to (\alpha,g,\beta)$ is an isomorphism onto $\mathfrak{J}(I) \times G \times \mathfrak{J}'(\mathrm{M})$. Going back to S it follows immediately that item i) is valid.

One frequently wishes to know when the projection homomorphism of $\Omega(S)$ into $\Lambda(S)$ (or $\mathrm{P}(S)$) is one-to-one. For in such a case, we may consider, e.g., $\tilde{\Lambda}(S)$ instead of $\Omega(S)$ which in some cases simplifies the calculations. To this end, we introduce a new concept.

V.3.13 DEFINITION. Let $S = \mathfrak{M}^0(I,G,\mathrm{M};P)$. The μ-th and ν-th rows of P are *left proportional* if there exists $c \in G$ such that $p_{\mu i} = cp_{\nu i}$ for all $i \in I$. The i-th and the j-th columns of P are *right proportional* if there exists $c \in G$ such that $p_{\mu i} = p_{\mu j}c$ for all $\mu \in M$.

V.3.14 THEOREM. The following conditions on $S = \mathfrak{M}^0(I,G,\mathrm{M};P)$ are equivalent.

i) For every left translation there exists at most one linked right translation.
ii) For every inner left translation there exists at most (exactly) one linked right translation.
iii) The left translation ι_S is linked to at most (exactly) one right translation.
iv) No two different rows of P are left proportional.
v) S is right reductive.

Proof. iii) *implies* iv). Suppose that $p_{\mu i} = cp_{\nu i}$ for some $c \in G$ and all $i \in I$. Define ψ and β on M by:

$$\mu\psi = c \quad \text{and} \quad \gamma\psi = 1 \quad \text{if} \quad \gamma \neq \mu,$$
$$\mu\beta = \nu \quad \text{and} \quad \gamma\beta = \gamma \quad \text{if} \quad \gamma \neq \mu.$$

It follows immediately that $p_{\gamma i} \neq 0$ if and only if $p_{(\gamma\beta)i} \neq 0$, and if so, then $(\gamma\psi)p_{(\mu\beta)i} = p_{\gamma i}$, which by 3.7 implies that (ι_I,φ_I) and (ψ,β) are linked, where φ_I maps I onto 1. By hypothesis, it then follows that $\psi = \psi_{\mathrm{M}}$, $\beta = \iota_{\mathrm{M}}$ since (i_I,φ_I) and $(\psi_{\mathrm{M}},\iota_{\mathrm{M}})$ are linked, where ψ_{M} maps M onto 1. But then $\mu = \mu\iota_{\mathrm{M}} = \mu\beta = \nu$.

iv) *implies* v). Suppose that for $a,b \neq 0$,

$$(i,a,\mu)(k,x,\gamma) = (j,b,\nu)(k,x,\gamma) \qquad ((k,x,\gamma) \in S). \tag{1}$$

For $p_{\mu k} \neq 0$ and $x = 1$, (1) yields $ap_{\mu k} = bp_{\nu k}$, whence we have $p_{\mu k} = a^{-1}bp_{\nu k}$. It also follows from (1) that $p_{\mu k} \neq 0$ if and only if $p_{\nu k} \neq 0$ so that $p_{\mu k} = a^{-1}bp_{\nu k}$ is valid for all $k \in I$. By iv) we have $\mu = \nu$, which together with $p_{\mu k} = a^{-1}bp_{\nu k}$ implies $a = b$. Since (1) also implies $i = j$, we have proved $(i,a,\mu) = (j,b,\nu)$.

The remaining implications follow from the next proposition.

V.3.15 PROPOSITION. Let S be any semigroup and let i), ii), iii) and v) be as in 3.14. Then i) implies ii) and iii), and v) implies i). Also ii) implies v) if S is weakly reductive and ii) implies i) if S is globally idempotent.

Proof. Exercise.

V.3.16. Exercises

1. Using the notation of the proof of 3.14, show that the following conditions on $S = \mathfrak{M}^0(I,G,M;P)$ are equivalent.
   i) If $\mathfrak{a}(\alpha,\varphi)$ is linked to both $(\psi,\beta)\mathfrak{b}$ and $(\psi,\beta')\mathfrak{b}$, then $\beta = \beta'$.
   ii) If ι_S is linked to (ψ_M,β) , then $\beta = \iota_M$.
   iii) $p_{\mu i} = p_{\nu i}$ for all $i \in I$ implies $\mu = \nu$.
   Also show that if $\mathfrak{a}(\alpha,\varphi)$ is linked to both $(\psi,\beta)\mathfrak{b}$ and $(\psi',\beta)\mathfrak{b}$, then $\psi = \psi'$.

2.* Let X be a nonempty set and by $\mathcal{P}(X)$ denote the set of all nonempty subsets of X. Define a $\mathcal{P}(X) \times X$-matrix P by: $P = (p_{Aa})$ where $p_{Aa} = 1$ if $a \in A$ and $p_{Aa} = 0$ if $a \notin A$, and let $S = \mathfrak{M}^0(X,1,\mathcal{P}(X);P)$. Setting $\mathfrak{F}_0(X) = \{\alpha \in \mathfrak{F}(X) \mid \textbf{rank}\ \alpha \leq 1\}$, prove: $S \cong \mathfrak{F}_0(X)$, $\Omega(S) \cong \Lambda(S) \cong \mathfrak{F}(X)$.

3. For $S = \mathfrak{M}^0(I,G,M;P)$, find the center of $\Omega(S)$.

4.* Show that $\mathfrak{F}(X)$ is a regular semigroup and find all its idempotents. Characterize completely regular elements of $\mathfrak{F}(X)$.

5. Find the idealizer of $\mathfrak{I}^0(X)$ in $\mathfrak{F}(X)$, and show that $\mathfrak{I}^0(X)$ is a left zero semigroup, a densely embedded ideal and the kernel of $\mathfrak{I}(X)$.

6. Find the center of $\mathfrak{I}(X)$ and of $\mathfrak{F}(X)$.

V.3.17 REFERENCES: Petrich [11], [12], [13], [14], [16], [17].

V.4. Inverse Semigroups

Along with the class of completely regular semigroups, the class of inverse semigroups represents the most important subclass of the class of regular semigroups. We consider here the translational hull of an arbitrary inverse semigroup; in the next two sections, we will give precise constructions for special kinds of inverse semigroups.

V.4.1 DEFINITION. An *inverse semigroup* is a regular semigroup in which every element has a unique inverse.

The unique inverse element of an element s of an inverse semigroup S will be denoted by s^{-1}. If s is also a completely regular element, this notation agrees with that introduced in IV.1.3. We will use the next lemma without express reference.

V.4.2 LEMMA. In any inverse semigroup S, we have

$$(xy)^{-1} = y^{-1}x^{-1}, \qquad (x^{-1})^{-1} = x, \qquad e^{-1} = e \qquad (x,y \in S,\ e \in E_S).$$

Proof. Exercise.

V.4.3 DEFINITION. A function φ mapping a semigroup S into a semigroup S' is an *antihomomorphism* if for all $a,b \in S$, $(ab)\varphi = (b\varphi)(a\varphi)$. The concepts of *antiisomorphism*, *antiendomorphism* and *antiautomorphism* have their obvious meaning.

For an inverse semigroup S, 4.2 says that the mapping $s \to s^{-1}$ $(s \in S)$ is an antiautomorphism of S leaving the idempotents of S fixed and whose square is the identity mapping. A kind of converse is provided by the following result.

V.4.4 LEMMA. A regular semigroup S having an antiendomorphism which leaves the idempotents of S fixed is an inverse semigroup.

Proof. Let φ be such an antiendomorphism, and let y and z be inverses of an element x in S. Then

$$y = yxy = y(xzx)y = (yx)\varphi(zx)\varphi y$$
$$= (zxyx)\varphi y = (zx)\varphi y = zxy$$

and analogously $z = zxy$ proving that $y = z$. Thus S is an inverse semigroup.

A result of general interest for regular semigroups is the following.

V.4.5 LEMMA. A regular semigroup S is an inverse semigroup if and only if its idempotents commute.

Proof. Let S be an inverse semigroup, e and f be idempotents of S, and $a = (ef)^{-1}$. A simple calculation shows that both ae and fa are also inverses of ef which by uniqueness of inverses implies $a = ae = fa$. But then $a^2 = (ae)(fa) = a(ef)a = a$, and thus $a = a^{-1} = ef$. Hence $ef \in E_S$ and analogously $fe \in E_S$. Consequently $(ef)(fe)(ef) = ef$ and $(fe)(ef)(fe) = fe$ proving that $(ef)^{-1} = fe$. On the other hand, $(ef)^{-1} = ef$ since $ef \in E_S$. Thus $ef = fe$ as required.

Conversely, suppose that idempotents of S commute, and let y and z be inverses of an element x of S. Then

$$y = yxy = y(xzx)y = y(xz)(xy) = y(xy)(xz) = yxz$$

and analogously $y = zxy$, which implies

$$y = yxy = (zxy)x(yxz) = z(xyxyx)z = zxz = z,$$

as required.

V.4.6 THEOREM. If S is an inverse semigroup, then so is $\Omega(S)$.

Proof. Let $(\lambda,\rho) \in \Omega(S)$ and define $\bar{\lambda}$ and $\bar{\rho}$ by:

$$\bar{\lambda}x = (x^{-1}\rho)^{-1}, \qquad x\bar{\rho} = (\lambda x^{-1})^{-1} \qquad (x \in S). \tag{1}$$

Then for any $x,y \in S$, we obtain

$$\bar{\lambda}(xy) = [(xy)^{-1}\rho]^{-1} = [(y^{-1}x^{-1})\rho]^{-1} = [y^{-1}(x^{-1}\rho)]^{-1}$$
$$= (x^{-1}\rho)^{-1}y = (\bar{\lambda}x)y,$$
$$(xy)\bar{\rho} = [\lambda(xy)^{-1}]^{-1} = [\lambda(y^{-1}x^{-1})]^{-1} = [(\lambda y^{-1})x^{-1}]^{-1}$$
$$= x(\lambda y^{-1})^{-1} = x(y\bar{\rho}),$$

$$x(\bar{\lambda}y) = x(y^{-1}\rho)^{-1} = [(y^{-1}\rho)x^{-1}]^{-1} = [y^{-1}(\lambda x^{-1})]^{-1}$$
$$= (\lambda x^{-1})^{-1}y = (x\bar{\rho})y$$

which proves that $(\bar{\lambda},\bar{\rho}) \in \Omega(S)$. From (1) follows

$$(\lambda x)^{-1} = x^{-1}\bar{\rho}, \qquad (x\rho)^{-1} = \bar{\lambda}x^{-1} \tag{2}$$

whence

$$\bar{\bar{\lambda}}x = (x^{-1}\bar{\rho})^{-1} = \lambda x, \qquad x\bar{\bar{\rho}} = \bar{\lambda}(x^{-1})^{-1} = x\rho,$$

and thus

$$\bar{\bar{\lambda}} = \lambda, \qquad \bar{\bar{\rho}} = \rho. \tag{3}$$

We then have, using (2),

$$[(\bar{\lambda}\lambda)(xx^{-1})]^2 = (\bar{\lambda}\lambda xx^{-1})(\bar{\lambda}\lambda xx^{-1}) = \bar{\lambda}[(\lambda xx^{-1})(\bar{\lambda}\lambda x)]x^{-1}$$
$$= \bar{\lambda}[(\lambda x)(x^{-1}\bar{\rho})(\lambda x)]x^{-1} = \bar{\lambda}[(\lambda x)(\lambda x)^{-1}(\lambda x)]x^{-1}$$
$$= \bar{\lambda}(\lambda x)x^{-1} = (\bar{\lambda}\lambda)(xx^{-1}) \tag{4}$$

which yields $[(\bar{\lambda}\lambda)(xx^{-1})]^{-1} = (\bar{\lambda}\lambda)(xx^{-1})$. Further, using (1), (2) and (3) we find

$$[(\bar{\lambda}\lambda)(xx^{-1})]^{-1} = \{\bar{\lambda}[\lambda(xx^{-1})]\}^{-1} = [\lambda(xx^{-1})]^{-1}\bar{\bar{\rho}}$$
$$= [\lambda(xx^{-1})]^{-1}\rho = [(xx^{-1})^{-1}\bar{\rho}]\rho = (xx^{-1})\bar{\rho}\rho.$$

Therefore

$$(\bar{\lambda}\lambda)(xx^{-1}) = (xx^{-1})(\bar{\rho}\rho), \tag{5}$$

and analogously,

$$(\lambda\bar{\lambda})(x^{-1}x) = (x^{-1}x)(\rho\bar{\rho}). \tag{6}$$

From (2) and (5), it follows that

$$(\lambda\bar{\lambda}\lambda)x = \lambda[(\bar{\lambda}\lambda)(xx^{-1})]x = \lambda[(xx^{-1})(\bar{\rho}\rho)]x$$
$$= (\lambda x)(x^{-1}\bar{\rho})(\lambda x) = (\lambda x)(\lambda x)^{-1}(\lambda x) = \lambda x.$$

Also from (2) and (6), we obtain

$$\begin{aligned} x(\rho\bar{\rho}\rho) &= x[(x^{-1}x)(\rho\bar{\rho})]\rho = x[(\lambda\bar{\lambda})(x^{-1}x)]\rho \\ &= (x\rho)(\bar{\lambda}x^{-1})(x\rho) = (x\rho)(x\rho)^{-1}(x\rho) = x\rho. \end{aligned}$$

Consequently

$$(\lambda,\rho)(\bar{\lambda},\bar{\rho})(\lambda,\rho) = (\lambda,\rho), \tag{7}$$

i.e., $\Omega(S)$ is regular. It may be shown analogously that

$$(\bar{\lambda},\bar{\rho})(\lambda,\rho)(\bar{\lambda},\bar{\rho}) = (\bar{\lambda},\bar{\rho}). \tag{8}$$

To complete the proof, by 4.4 it suffices to show that the mapping

$$\theta:(\lambda,\rho) \to (\bar{\lambda},\bar{\rho}) \qquad ((\lambda,\rho) \in \Omega(S))$$

is an antiendomorphism leaving the idempotents of $\Omega(S)$ fixed. Indeed, for any $(\varphi,\psi) \in \Omega(S)$, we have

$$\bar{\lambda}\bar{\varphi}x = \bar{\lambda}(x^{-1}\psi)^{-1} = (x^{-1}\psi\rho)^{-1} = \overline{\varphi\lambda}x$$

and dually $\bar{\rho}\bar{\psi} = \overline{\psi\rho}$ so that θ is an antiendomorphism. Finally let (λ,ρ) be an idempotent of $\Omega(S)$. Then for any $x \in S$,

$$\begin{aligned} \bar{\lambda}x &= \bar{\lambda}\lambda\bar{\lambda}x && \text{by (8)} \\ &= (\bar{\lambda}\lambda)[\overline{\bar{\lambda}\lambda}(xx^{-1})]x && \text{by } \lambda^2 = \lambda \text{ and (3)} \\ &= (\bar{\lambda}\lambda)[(xx^{-1})\bar{\bar{\rho}}\bar{\rho}](xx^{-1})x && \text{by (5)} \\ &= [\bar{\lambda}\lambda(xx^{-1})][\overline{\bar{\lambda}\lambda}(xx^{-1})]x && \text{by the linking condition} \\ &= [\lambda\bar{\lambda}(xx^{-1})][\bar{\lambda}\lambda(xx^{-1})]x && \text{by (3), (4) and 4.5} \\ &= \lambda\bar{\lambda}[(xx^{-1})(\bar{\lambda}\lambda x)] && \\ &= \lambda\bar{\lambda}[(xx^{-1})\bar{\rho}\rho]x && \text{by the linking condition} \\ &= (\lambda\bar{\lambda}^2\lambda)x && \text{by (5)} \\ &= \lambda\bar{\lambda}\lambda x && \text{since } \rho^2 = \rho \text{ implies } \bar{\lambda}^2 = \bar{\lambda} \\ &= \lambda x && \text{by (7).} \end{aligned}$$

One shows similarly that $\bar{\rho} = \rho$ proving that θ leaves the idempotents of $\Omega(S)$ fixed.

V.4.7. Exercises

1. Let T be a nonempty subset of an inverse semigroup S. Show that $\{x_1x_2 \ldots x_n |$ either $x_i \in T$ or $x_i^{-1} \in T\}$ is the least inverse subsemigroup of S containing T.

2. Let S be an inverse semigroup. Show that if $\lambda \in \Lambda(S)$ maps idempotents onto idempotents, then $\lambda^2 = \lambda$. Also prove that the following conditions on $(\lambda,\rho) \in \Omega(S)$ are equivalent: (i) $\lambda^2 = \lambda$, (ii) $\rho^2 = \rho$, (iii) λ maps idempotents onto idempotents, (iv) ρ maps idempotents onto idempotents. Deduce that $E_{\Omega(S)} \cong \Omega(E_S)$.

3. Show that every inverse semigroup is reductive.

4. Let S be an inverse semigroup. Show that an automorphism of $\Omega(S)$ which leaves the idempotents of $\Pi(S)$ fixed also leaves all idempotents of $\Omega(S)$ fixed. Deduce that if $\theta \in \mathfrak{A}(S)$ leaves the idempotents of S fixed, then $\bar{\theta}$ leaves the idempotents of $\Omega(S)$ fixed.

5. Show that for an inverse semigroup S without identity, $\Omega(E_S) \cong E_S^1$ if and only if $\Omega(S) = \Pi(S) \cup \Sigma(S)$.

6. Show that a regular semigroup S has an antiendomorphism θ such that for every $s \in S$, $s(s\theta)$ is a right identity of S if and only if S is a left group.

7. Show that in an inverse semigroup S, $i_S(E_S) = E_S$ and

$$\{x \in S \mid xe = ex \text{ for all } e \in E_S\}$$

is the greatest completely regular subsemigroup of S containing E_S.

8. Let θ be an antiisomorphism of a semigroup S onto a semigroup T. For every $(\lambda,\rho) \in \Omega(S)$ define $\bar{\lambda}$ and $\bar{\rho}$ by:

$$\bar{\lambda}t = [(t\theta^{-1})\rho]\theta, \qquad t\bar{\rho} = [\lambda(t\theta^{-1})]\theta \qquad (t \in T).$$

What kind of mapping is $(\lambda,\rho) \to (\bar{\lambda},\bar{\rho})$?

9. Show that a right cancellative inverse semigroup must be a group.

10. Let V be an extension of a semigroup S by a semigroup Q with zero. Let P stand for any of the following properties of semigroups: (i) regular, (ii) completely regular, (iii) inverse semigroup, (iv) semilattice. Show that V has property P if and only if both S and Q have property P.

11. Let I be a nonempty set and G be a group. Define

$$\chi : \mathfrak{F}(I)\ w\ell\ G \to G\ wr\ \mathfrak{F}'(I)$$

by $\chi:(\alpha,\varphi) \to (\psi,\beta)$ where α and β and also φ and ψ are the same functions written on opposite sides of the argument. Show that χ is an antiisomorphism of $\mathfrak{F}(I)\ w\ell\ G$ onto $G\ wr\ \mathfrak{F}'(I)$ if and only if G is commutative.

12. Show that a regular semigroup S is an inverse semigroup if and only if for any $a,x,y \in S$, $a = axa = aya$ implies $xax = yay$.

V.4.8 REFERENCES: Ponizovskiĭ [1].

V.5. Brandt Semigroups

We will now characterize the translational hull of a Rees matrix semigroup which is also an inverse semigroup. It is convenient to first describe such semigroups in a suitable manner.

V.5.1 PROPOSITION. The following conditions on $S = \mathfrak{M}^0(I,G,M;P)$ are equivalent.

i) S is an inverse semigroup.
ii) Every row and every column of P contains exactly one nonzero entry.
iii) $S \cong \mathfrak{M}^0(I,G,I;\Delta)$ where Δ is the $I \times I$ identity matrix.
iv) The product of any two distinct idempotents equals zero.

Proof. i) *implies* ii). If $p_{\lambda i} \neq 0$ and $p_{\lambda j} \neq 0$, then both $(i,p_{\lambda i}^{-1},\lambda)$ and $(j,p_{\lambda j}^{-1},\lambda)$ are inverses of $(i,p_{\lambda i}^{-1},\lambda)$, so $(i,p_{\lambda i}^{-1},\lambda) = (j,p_{\lambda j}^{-1},\lambda)$ which implies $i = j$. Similarly $p_{\lambda i} \neq 0$ and $p_{\mu i} \neq 0$ imply $\lambda = \mu$, proving that P has the required property.

ii) *implies* iii). In view of the hypothesis, we may define a one-to-one function φ mapping M onto I by the requirement $p_{\mu(\mu\varphi)} \neq 0$. We further define a function χ from S into $\mathfrak{M}^0(I,G,I;\Delta)$ by

$$\chi:(i,a,\mu) \to (i,{p^{-1}}_{(i\varphi^{-1})i}\,a{p^{2}}_{\mu(\mu\varphi)},\mu\varphi) \qquad ((i,a,\mu) \in S).$$

Then

$$
\begin{aligned}
(i,a,\mu)\chi(j,b,\nu)\chi &= (i,p^{-1}{}_{(i\varphi^{-1}),i}ap^{2}{}_{\mu(\mu\varphi)},\mu\varphi)(j,p^{-1}{}_{(j\varphi^{-1}),j}bp^{2}{}_{\nu(\nu\varphi)},\nu\varphi) \\
&= \begin{cases} (i,p^{-1}{}_{(i\varphi^{-1}),i}ap^{2}{}_{\mu(\mu\varphi)}p^{-1}{}_{(j\varphi^{-1}),j}bp^{2}{}_{\nu(\nu\varphi)},\nu\varphi) & \text{if } \mu\varphi = j \\ 0 & \text{otherwise} \end{cases} \\
&= \begin{cases} (i,p^{-1}{}_{(i\varphi^{-1}),i}ap_{\mu j}bp^{2}{}_{\nu(\nu\varphi)},\nu\varphi) & \text{if } \mu\varphi = j \\ 0 & \text{otherwise} \end{cases} \\
&= (i,ap_{\mu j}b,\nu)\chi = [(i,a,\mu)(j,b,\nu)]\chi
\end{aligned}
$$

proving that χ is a homomorphism. The proof that χ is one-to-one and onto consists of a simple verification and is left as an exercise.

iii) *implies* iv). It is an easy calculation to show that the isomorphic copy $\mathfrak{M}^0(I,G,I;\Delta)$ enjoys the desired property.

iv) *implies* i). This follows immediately from 4.5.

V.5.2 DEFINITION. A semigroup S isomorphic to a semigroup of the form $\mathfrak{M}^0(I,G,I;\Delta)$ is a *Brandt semigroup.*

Hence to find the translational hull of a Brandt semigroup, it suffices to consider a semigroup $S = \mathfrak{M}^0(I,G,I;\Delta)$. On the one hand, in V.3 we have constructed the translational hull of an arbitrary Rees matrix semigroup, and on the other, from 4.6, we know that for our S, $\Omega(S)$ is an inverse semigroup.

V.5.3 DEFINITION. Let X be a nonempty set. A *one-to-one partial transformation* on X is a partial transformation on X which is one-to-one on its domain. The *set* $\mathcal{I}(X)$ *of all one-to-one partial transformations on* X written on the left, is easily seen to be a subsemigroup of $\mathcal{F}(X)$; one defines $\mathcal{I}'(X)$ dually. Note that $\varnothing$ is an element both of $\mathcal{I}(X)$ and $\mathcal{I}'(X)$.

If $\alpha \in \mathcal{I}(X)$, let α^{-1} be the unique function for which $\mathbf{d}\alpha^{-1} = \mathbf{r}\alpha$, $\mathbf{r}\alpha^{-1} = \mathbf{d}\alpha$, $\alpha\alpha^{-1} = \iota_{\mathbf{r}\alpha}$, $\alpha^{-1}\alpha = \iota_{\mathbf{d}\alpha}$. Similarly for $\alpha \in \mathcal{I}'(X)$. It is easy to verify that $\mathcal{I}(X)$ is an inverse semigroup with α^{-1} the unique inverse of α.

V.5.4 PROPOSITION. For $S = \mathfrak{M}^0(I,G,I;\Delta)$, we have

i) $\tilde{\Lambda}(S) = \{\lambda \in \Lambda(S) \mid \mathfrak{a}\lambda = (\alpha,\varphi), \alpha \in \mathcal{I}(I)\} \cong \mathcal{I}(I)\ w\ell\ G$,
ii) $\tilde{\mathrm{P}}(S) = \{\rho \in \mathrm{P}(S) \mid \rho\mathfrak{b} = (\psi,\beta), \beta \in \mathcal{I}'(I)\} \cong G\ wr\ \mathcal{I}'(I)$,
iii) $\Omega(S) \cong \tilde{\Lambda}(S) \cong \tilde{\mathrm{P}}(S)$.

Proof. To simplify the arguments, we will identify λ with $\mathfrak{a}\lambda$, ρ with $\rho\mathfrak{b}$. Let $((\alpha,\varphi), (\beta,\psi)) \in \Omega(S)$. Then 3.7 i) becomes $i \in \mathbf{d}\alpha$, $j = \alpha i$ if and only if $j \in \mathbf{d}\beta$, $j\beta = i$. Consequently $\beta = \alpha^{-1}$ proving that $\alpha \in \mathcal{I}(I)$. Conversely, if $(\alpha,\varphi) \in \Lambda(S)$ with $\alpha \in \mathcal{I}(I)$, then letting $j\beta = \alpha^{-1}j$ and $j\psi = \varphi\alpha^{-1}j$ for all $j \in \mathbf{r}\alpha$, by 3.7 we conclude that $((\alpha,\varphi), (\psi,\beta)) \in \Omega(S)$. This proves the equality in i); the required isomorphism in i) is the restriction of $\mathfrak{a}$ to $\tilde{\Lambda}(S)$ (see 3.4). Item ii) is proved analogously; item iii) follows from 3.14.

Let $\mathcal{I}_0(X)$ be the subset (in fact, ideal) of $\mathcal{I}(X)$ consisting of those elements α for which $|\mathbf{d}\alpha| = |\mathbf{r}\alpha| \leq 1$, and let 1 denote a one-element group. Then we obtain the following characterization of $\mathcal{I}(X)$.

V.5.5 COROLLARY. For $S = \mathfrak{M}^0(I,1,I;\Delta)$, we have

$$\Omega(S) \cong \mathcal{I}(I), \qquad \Pi(S) \cong \mathcal{I}_0(I).$$

Proof. Exercise.

V.5.6 LEMMA. Let $T = \mathcal{I}(I)\ w\ell\ G$; then the group of units of T is given by

$$\{(\alpha,\varphi) \in T \mid \alpha \in \mathcal{S}(I)\} \cong \mathcal{S}(I)\ w\ell\ G$$

where $\mathcal{S}(I)$ is the *symmetric group on* I, and the center of T is given by

$$\{(\iota_I,\varphi_c) \mid c \in \mathcal{C}(G)\} \cup 0 \cong \mathcal{C}(G^0)$$

where $\varphi_c i = c$ for all $i \in I$.

Proof. The proof of the first statement is left as an exercise. Let $(\alpha,\varphi) \in \mathcal{C}(T)$. Then for all $(\xi,\eta) \in T$, we have $(\alpha,\varphi)(\xi,\eta) = (\xi,\eta)(\alpha,\varphi)$ which is equivalent to $\alpha\xi = \xi\alpha$ and $\varphi^{\xi}\cdot\eta = \eta^{\alpha}\cdot\varphi$. In particular, let $\mathbf{d}\xi = \mathbf{r}\xi = i$. Then $i \in \mathbf{d}\alpha$, let $j = \alpha i$; so $j = \alpha i = \alpha(\xi i) = \xi\alpha i = \xi j$ which implies $j = i$. Consequently $\alpha = \iota_I$. The second equation now yields $(\varphi\xi i)(\eta i) = (\eta i)(\varphi i)$ for all $i \in \mathbf{d}\xi$. Take $i,j \in I$ to be arbitrary; let $\mathbf{d}\xi = i$, $\xi i = j$, $\eta i = 1$. Then $\varphi j = \varphi i$ proving that φ is a constant, say $\varphi = \varphi_c$. We then obtain $c(\eta i) = (\eta i)c$, and since ηi is arbitrary, we must have $c \in \mathcal{C}(G)$. Conversely, it is easy to see that $(\iota_I,\varphi_c) \in \mathcal{C}(T)$ if $c \in \mathcal{C}(G)$.

The next theorem provides an example announced after 2.9.

V.5.7 THEOREM. For $S = \mathfrak{M}^0(I,G,I;\Delta)$, we have

$$\Sigma(S) \cap \mathcal{C}(\Omega(S)) = \mathcal{C}(\Sigma(S))$$

except when $|I| = 2$ and $|G| = 1$.

Proof. In view of 5.5 and 5.4, we may substitute $\Sigma(S)$ by $\mathcal{S}(I)\ w\ell\ G$ and $\Omega(S)$ by $\mathcal{J}(I)\ w\ell\ G$. Moreover, 5.6 gives the center of $\mathcal{J}(I)\ w\ell\ G$, while it is well known that $\mathcal{C}(\mathcal{S}(I))$ is trivial if and only if $|I| \neq 2$. Hence the problem reduces to proving that

$$\{(\iota_I,\varphi_c) \mid c \in \mathcal{C}(G)\} = \mathcal{C}(\mathcal{S}(I)\ w\ell\ G) \tag{1}$$

except when $|I| = 2$ and $|G| = 1$. If $|I| = 2$ and $|G| = 1$, then the left hand side of (1) has only one element, while the right hand side has two elements.

We will show next that if $|I| = 2$ or $|G| = 1$ fails, then (1) holds. It suffices to show that the right hand side is contained in the left hand side. Let $(\alpha,\varphi) \in \mathcal{C}(\mathcal{S}(I)\ w\ell\ G)$. Then for any $(\xi,\eta) \in \mathcal{S}(I)\ w\ell\ G$, we have $(\alpha,\varphi)(\xi,\eta) = (\xi,\eta)(\alpha,\varphi)$ which implies $\alpha\xi = \xi\alpha$ and $\varphi^{\xi}\cdot\eta = \eta^{\alpha}\cdot\varphi$. It follows that

$$\alpha \in \mathcal{C}(\mathcal{S}(I)), \qquad (\varphi\xi i)(\eta i) = (\eta\alpha i)(\varphi i) \qquad (i \in I). \tag{2}$$

If $|I| > 2$, then $\alpha \in \mathcal{C}(\mathcal{S}(I))$ implies $\alpha = \iota_I$, which by (2) implies $(\varphi\xi i)(\eta i) = (\eta i)(\varphi i)$. As in the proof of 5.6, one shows easily that $\varphi = \varphi_c$ for some $c \in \mathcal{C}(G)$. Hence $(\alpha,\varphi) = (\iota_I,\varphi_c)$ with $c \in \mathcal{C}(G)$ and (1) holds.

The case in which $|I| = 1$ is trivial, so we may now consider the only remaining case in which $|I| = 2$ and $|G| > 1$. Let $I = \{1,2\}$, e be the identity of G, and $\theta = \begin{pmatrix} 1 & 2 \\ 2 & 1 \end{pmatrix}$. Let $(\theta,\tau) \in \mathcal{S}(I)\ w\ell\ G$, and define $\varphi: I \to G$ by $\varphi 1 = e$, $\varphi 2 \neq (\tau 2)(\tau 1)^{-1}$, otherwise arbitrary. This is possible since G has more than one element. Thus $(\tau\theta 1)(\varphi 1) = \tau 2$ and $(\varphi\theta 1)(\tau 1) = (\varphi 2)(\tau 1)$; these two are different in view of the fact that $\varphi 2 \neq (\tau 2)(\tau 1)^{-1}$. But then $(\theta,\tau)(\theta,\varphi) \neq (\theta,\varphi)(\theta,\tau)$, so (θ,τ) can not be in the center. Hence every element in the center is of the form (ι_I,τ) for some τ. That $\tau = \varphi_c$ for $c \in \mathcal{C}(G)$ is proved similarly as in the proof of 5.6. Hence (1) holds in this case also.

V.5.8. Exercises

1. Let S be an inverse subsemigroup of $\mathcal{I}(X)$ for some nonempty set X, not containing $\varnothing$. On S define a relation τ by:

$$\alpha \tau \beta \quad \text{if there exists} \quad \gamma \in S \quad \text{such that} \quad \mathbf{d}\gamma \subseteq \mathbf{d}\alpha \cap \mathbf{d}\beta \quad \text{and} \quad \alpha|_{\mathbf{d}\gamma} = \gamma = \beta|_{\mathbf{d}\gamma}.$$

   Show that τ is the least group congruence on S.

2. For a nonempty set I and a group G, let $T = \mathcal{I}(I)\ w\ell\ G$. Show that a nonzero element (α,φ) of T is idempotent if and only if $\alpha i = i$ and $\varphi i = 1$ for all $i \in \mathbf{d}\alpha$, and that in such a case

$$G_{(\alpha,\varphi)} = \{(\xi,\eta) \in T \mid \mathbf{d}\xi = \mathbf{d}\alpha, \quad \xi \in \mathcal{S}(\mathbf{d}\alpha)\} \cong \mathcal{S}(\mathbf{d}\alpha)\ w\ell\ G.$$

3. Let X be a nonempty set. Find the idealizer of $\mathcal{I}_0(X)$ in $\mathcal{F}(X)$. Show that $\mathcal{I}_0(X)$ is a densely embedded ideal of $\mathcal{I}(X)$ contained in all nonzero ideals of $\mathcal{I}(X)$.

4. Let $S = \mathcal{M}^0(I,G,I;\Delta)$, and for any nonempty subset J of I, let $\mathcal{K}_J$ denote the set of all subsets K of S^* such that for every $j \in J$ there exists a unique element $(j,a,i) \in K$ and a unique element $(k,b,j) \in K$, with the multiplication of subsets. Show that for a nonzero idempotent $(\alpha,\varphi) \in \tilde{\Lambda}(S)$, $G_{(\alpha,\varphi)} \cong \mathcal{K}_{\mathbf{d}\alpha}$. Deduce that $\Sigma(S) \cong \mathcal{K}_I \cong \mathcal{S}(I)\ w\ell\ G$.

5. Show that the set of all automorphisms and antiautomorphisms of a semigroup S forms a group having the group of automorphisms as a subgroup of index 1 or 2 (and thus a normal subgroup).

V.5.9 REFERENCES: Brandt [1], Lallement and Petrich [2], Petrich [12], Warne [1], [2].

V.6. Semilattices of Groups

For a semilattice of groups S given in the form $[Y;G_\alpha,\psi_{\alpha,\beta}]$, we will now construct two isomorphic copies of $\Omega(S)$; one of these is a generalization of the inverse limit of a directed set of groups and the other one represents $\Omega(S)$ in the form of a semilattice of groups. Note that a semigroup S is a semilattice of groups if and only if S is an inverse completely regular semigroup, so semilattices of groups represent the intersection of these two important classes of regular semigroups. We begin with semilattices.

V.6.1 LEMMA. Let Y be a semilattice. Then left translations of Y coincide with I-endomorphisms of Y where I is an ideal of Y. The set $\mathcal{R}_Y$ of retract ideals coincides with the set of all ideals I having the property: For every $\alpha \in Y$, $I \cap J(\alpha)$ is a principal ideal. The mapping χ defined by:

$$\chi:\lambda \rightarrow \lambda Y \qquad (\lambda \in \Lambda(Y))$$

is an isomorphism of $\Lambda(Y)$ onto $\mathcal{R}_Y$ where the multiplication in the latter is the set theoretical intersection. Moreover, for $I \in \mathcal{R}_Y$ the corresponding left translation is given by: $I \cap J(\alpha) = J(\lambda\alpha)$ for every $\alpha \in Y$.

Proof. If $\lambda \in \Lambda(Y)$, then for any $\alpha \in Y$, we have

$$\lambda\alpha = (\lambda\alpha)^2 = (\lambda\alpha)(\lambda\alpha) = \lambda[(\lambda\alpha)\alpha] = \lambda^2\alpha,$$

and hence for any $\alpha,\beta \in Y$,

$$\lambda(\alpha\beta) = \lambda^2(\alpha\beta) = \lambda[(\lambda\alpha)\beta] = (\lambda\alpha)(\lambda\beta).$$

This shows that λ is an endomorphism and that λ fixes every element of λY; it is clear that λY is an ideal of Y. Conversely, let λ be an I-endomorphism of Y for some ideal I. Then for any $\alpha,\beta \in Y$, we obtain

$$(\lambda\alpha)\beta = \lambda[(\lambda\alpha)\beta] = (\lambda^2\alpha)(\lambda\beta) = (\lambda\alpha)(\lambda\beta) = \lambda(\alpha\beta)$$

since $(\lambda\alpha)\beta \in I$ and is thus fixed by λ. The second statement of the lemma follows easily from III.4.7. Let $\lambda,\lambda' \in \Lambda(S)$ and let $\alpha \in Y$. Then

$$\begin{aligned}\lambda\lambda'\alpha &= \lambda[(\lambda'\alpha)\alpha] = \lambda[\alpha(\lambda'\alpha)] = (\lambda\alpha)(\lambda'\alpha)\\ &= (\lambda'\alpha)(\lambda\alpha) = \lambda'[\alpha(\lambda\alpha)] \in \lambda' Y\end{aligned}$$

so that $\lambda\lambda' Y \subseteq \lambda Y \cap \lambda' Y$; conversely, if $\alpha = \lambda\beta = \lambda'\gamma$, then

$$\alpha = (\lambda\beta)(\lambda'\gamma) = \lambda[\beta(\lambda'\gamma)] = \lambda[(\lambda'\beta)\gamma] = \lambda\lambda'(\beta\gamma) \in \lambda\lambda' Y$$

and thus $\lambda Y \cap \lambda' Y \subseteq \lambda\lambda' Y$. Consequently $\lambda Y \cap \lambda' Y = \lambda\lambda' Y$ and χ is a homomorphism. That χ is one-to-one follows from the statements already proved and III.4.4. It is now clear that χ maps $\Lambda(S)$ onto $\mathcal{R}_Y$. We know already that for a given $I \in \mathcal{R}_Y$, a function λ can be defined on Y by the condition: $I \cap J(\alpha) = J(\lambda\alpha)$. Then $\lambda\alpha$ is the greatest element in I with the property of being less than or equal to α. Hence the proof of III.4.7 quickly implies that λ is an I-endomorphism of Y and hence a left translation.

V.6.2 COROLLARY. If Y is a semilattice, so is $\Omega(Y)$.

Proof. This follows from 6.1 and also from III.5.17.

Note that the isomorphism χ in 6.1 maps the set of all inner left translations of Y onto the set of all principal ideals of Y. We now fix a semilattice of groups $S = [Y; G_\alpha, \psi_{\alpha,\beta}]$, and first introduce the following notation. Following the usage in group theory, we write

$$inv\ lim\ \{G_\alpha\}_{\alpha \in Y} = \{(a_\alpha)_{\alpha \in Y} \in \prod_{\alpha \in Y} G_\alpha \mid a_\alpha \psi_{\alpha,\beta} = a_\beta \quad \text{if} \quad \alpha > \beta\}.$$

As above, $\mathcal{R}_Y$ denotes the semigroup of all retract ideals of Y; let

$$\mathcal{R}_S = \bigcup_{I \in R_Y} inv\ lim\ \{G_\alpha\}_{\alpha \in I}$$

with the multiplication

$$(a_\alpha)_{\alpha \in I} \cdot (b_\beta)_{\beta \in J} = (a_\gamma b_\gamma)_{\gamma \in I \cap J}$$

This induces the coordinatewise multiplication on each inv lim $\{G_\alpha\}_{\alpha \in I}$. We further denote by e_α the identity of G_α, and by $\mathcal{P}_Y$ the semigroup of all principal ideals of Y.

V.6.3 PROPOSITION. Each *inv lim* $\{G_\alpha\}_{\alpha \in I}$ is a group and

$$\mathcal{R}_S \cong [\mathcal{R}_Y;\ inv\ lim\ \{G_\alpha\}_{\alpha \in I},\ \varphi_{I,J}],$$

where

$$(a_\alpha)_{\alpha \in I} \varphi_{I,J} = (a_\alpha)_{\alpha \in J} \quad \text{if} \quad I \supseteq J.$$

Proof. The first statement is easy to verify. From the very construction, it is clear that $\mathcal{R}_S$ is a semilattice of groups *inv lim* $\{G_\alpha\}_{\alpha \in I}$. For $I \supseteq J$, we have

$$(a_\alpha)_{\alpha \in I}\varphi_{I,J} = (a_\alpha)_{\alpha \in J} = (a_\alpha)_{\alpha \in I} \cdot (e_\alpha)_{\alpha \in J}$$

where $(e_\alpha)_{\alpha \in J}$ is the identity of *inv lim* $\{G_\alpha\}_{\alpha \in J}$. It then follows that the homomorphisms $\varphi_{I,J}$ determine the multiplication in $\mathcal{R}_S$. A detailed proof is left as an exercise.

The principal result of this section can now be proved.

V.6.4 THEOREM. For $S = [Y; G_\alpha, \psi_{\alpha,\beta}]$ where each G_α is a group, the function $\mathfrak{r}$ defined on $\Lambda(S)$ by

$$\mathfrak{r}: \lambda \to (\lambda e_\alpha)_{\alpha \in I} \quad \text{where} \quad I = \{\alpha \in Y \mid \lambda S \cap G_\alpha \neq \varnothing\} \tag{1}$$

is an isomorphism of $\Lambda(S)$ onto $\mathcal{R}_S$. Furthermore

$$\mathfrak{r}\Gamma(S) = \{(a_\alpha)_{\alpha \in I} \in \mathcal{R}_S \mid I \in \mathcal{P}_Y\}.$$

Proof. Let $\lambda \in \Lambda(S)$. For any elements $x,y \in G_\alpha$, we have $\lambda x \in G_\gamma$ and $\lambda y \in G_\delta$ for some $\gamma, \delta \in Y$. Then $\lambda x = \lambda(yy^{-1}x) = (\lambda y)y^{-1}x$ implies $\gamma \leq \delta$; by symmetry we also have $\delta \leq \gamma$ so that $\gamma = \delta$. Denoting by $x \to \bar{x}$ the natural homomorphism of S onto Y, we thus may define a function $\bar{\lambda}$ on Y by $\bar{\lambda}\bar{x} = \overline{\lambda x}$ $(x \in S)$. Consequently

$$\bar{\lambda}(\bar{x}\bar{y}) = \bar{\lambda}\overline{xy} = \overline{\lambda(xy)} = \overline{(\lambda x)y} = \overline{\lambda x}\bar{y} = (\bar{\lambda}\bar{x})\bar{y}$$

which shows that $\bar{\lambda} \in \Lambda(Y)$. According to 6.1, $\bar{\lambda}$ is completely determined by the retract ideal $I = \bar{\lambda}Y$, and in addition, I coincides with the ideal I defined in (1). It follows that $\lambda e_\alpha \in G_\alpha$ for all $\alpha \in I$, and denoting the multiplication in S by $*$, for $\alpha > \beta$ we have

$$\lambda e_\beta = \lambda(e_\alpha * e_\beta) = (\lambda e_\alpha) * e_\beta = (\lambda e_\alpha)\psi_{\alpha,\beta}$$

which shows that $\mathfrak{r}$ maps $\Lambda(S)$ into $\mathcal{R}_S$.

Now let $\lambda, \mu \in \Lambda(S)$. With the notation introduced above, we have $\bar{\lambda}, \bar{\mu} \in \Lambda(Y)$ and by 6.1, $(\bar{\lambda}\bar{\mu})Y = \bar{\lambda}Y \cap \bar{\mu}Y$. Letting $\mathfrak{r}\lambda = (c_\alpha)_{\alpha \in I}$ and $\mu = (d_\alpha)_{\alpha \in J}$, we obtain for all $e_\alpha \in \lambda\mu S$,

$$(\lambda\mu)e_\alpha = \lambda(\mu e_\alpha) = \lambda d_\alpha = \lambda(e_\alpha * d_\alpha) = (\lambda e_\alpha) * d_\alpha = c_\alpha * d_\alpha$$

which shows that $\mathfrak{r}$ is a homomorphism.

Suppose next that $\mathfrak{r}\lambda = \mathfrak{r}\mu = (c_\alpha)_{\alpha \in I}$. By 6.1 we have, $I \cap J(a) = J(\bar{\lambda}\alpha) = J(\bar{\mu}\alpha)$ for all $\alpha \in Y$. Hence for any $a \in G_\alpha$, we have

$$\begin{aligned}\lambda a &= \lambda(e_\alpha * a) = (\lambda e_\alpha) * a = (\lambda e_{\bar{\lambda}\alpha}) * a = c_{\bar{\lambda}\alpha} * a \\ &= (\mu e_{\bar{\mu}\alpha}) * a = (\mu e_\alpha) * a = \mu(e_\alpha * a) = \mu a\end{aligned}$$

which proves that $\mathfrak{r}$ is one-to-one.

Let $(c_\gamma)_{\gamma \in I} \in \mathfrak{R}_S$. On S define a function λ by:

$$\lambda a = c_\gamma * a \quad \text{if} \quad a \in G_\alpha \quad \text{and} \quad I \cap J(\alpha) = J(\gamma). \tag{2}$$

Since I is a retract ideal of Y, 6.1 implies that for every $\alpha \in Y$ there is a unique γ satisfying (2). It is clear that this λ also induces a function $\bar{\lambda}$ on Y the same way as the left translation λ considered above, and that $I \cap J(\alpha) = J(\bar{\lambda}\alpha)$ for all $\alpha \in Y$. Hence 6.1 implies $\bar{\lambda} \in \Lambda(Y)$, and we have $\lambda a = c_{\bar{\lambda}\alpha} * a$ if $a \in G_\alpha$. For any $a \in G_\alpha$, $b \in G_\beta$, we compute

$$\begin{aligned}(\lambda a) * b &= (c_{\bar{\lambda}\alpha} * a) * b = (c_{\bar{\lambda}\alpha}\psi_{\bar{\lambda}\alpha,(\bar{\lambda}\alpha)\beta})(a\psi_{\alpha,(\bar{\lambda}\alpha)\beta})(b\psi_{\beta,(\bar{\lambda}\alpha)\beta}) \\ &= c_{(\bar{\lambda}\alpha)\beta}[(a\psi_{\alpha,\alpha\beta})(b\psi_{\beta,\alpha\beta})]\psi_{\alpha\beta,(\bar{\lambda}\alpha)\beta} \\ &= c_{\bar{\lambda}(\alpha\beta)} * (a * b) = \lambda(a * b).\end{aligned}$$

Consequently $\lambda \in \Lambda(S)$ and it is clear that $\mathfrak{r}\lambda = (c_\gamma)_{\gamma \in I}$, proving that σ maps $\Lambda(S)$ onto $\mathfrak{R}_S$.

If $a \in G_\alpha$, then $\overline{\lambda_a} = J(\alpha)$. Conversely, if $(a_\beta)_{\beta \in J(\alpha)} \in \mathfrak{R}_S$, then for any $\beta \in J(\alpha)$, we have

$$\begin{aligned}a_\beta &= a_\alpha\psi_{\alpha,\beta} = a_\alpha * e_\beta = a_\alpha e_\alpha * e_\beta = (\lambda_{a_\alpha} e_\alpha) * e_\beta \\ &= \lambda_{a_\alpha}(e_\alpha * e_\beta) = \lambda_{a_\alpha} e_\beta\end{aligned}$$

which implies $\mathfrak{r}\lambda_{a_\alpha} = (a_\beta)_{\beta \in J(\alpha)}$. This proves the last assertion of the theorem.

V.6.5 COROLLARY. With the notation of 6.4, we have

i) $\mathrm{P}(S) \cong \Omega(S) \cong \Lambda(S) \cong \mathfrak{R}_S$,
ii) $\Delta(S) \cong \Pi(S) \cong \Gamma(S) \cong S$,
iii) $\Sigma(S) \cong$ *inv lim* $\{G_\alpha\}_{\alpha \in Y}$.

Proof. First note that a semilattice of groups is reductive, which insures that the projections of $\Omega(S)$ into both $\Lambda(S)$ and $\mathrm{P}(S)$ are one-to-one. Hence to establish the third isomorphism in part i), in view of 6.4, it suffices to show that $\tilde{\Lambda}(S) = \Lambda(S)$. Thus let $\lambda \in \Lambda(S)$. Letting $\mathfrak{r}\lambda = (c_\gamma)_{\gamma \in I}$, we know by the proof of 6.4 that λ can be given as in (2) of 6.4. We now define a function ρ in a sense dualizing (2) of 6.4 as follows.

$$a\rho = a * c_\gamma \quad \text{if} \quad a \in G_\alpha \quad \text{and} \quad I \cap J(\alpha) = J(\gamma).$$

A proof symmetric to the one in 6.4 shows that $\rho \in P(S)$. To show that λ and ρ are linked, we let $a \in G_\alpha$, $b \in G_\beta$, $I \cap J(\alpha) = J(\gamma)$, $I \cap J(\beta) = J(\delta)$. Then

$$\begin{aligned} J(\alpha\delta) &= J(\alpha) \cap J(\delta) = J(\alpha) \cap (I \cap J(\beta)) = (I \cap J(\alpha)) \cap J(\beta) \\ &= J(\gamma) \cap J(\beta) = J(\gamma\beta) \end{aligned}$$

so that $\alpha\delta = \gamma\beta$. It follows that $c_\delta\psi_{\delta,\alpha\delta} = c_\gamma\psi_{\gamma,\gamma\beta}$, since $(c_\gamma)_{\gamma \in I} \in \mathfrak{R}_S$, whence

$$\begin{aligned} a * (\lambda b) &= a * c_\delta * b = (a\psi_{\alpha,\alpha\delta})(c_\delta\psi_{\delta,\alpha\delta})(b\psi_{\beta,\alpha\delta}) \\ &= (a\psi_{\alpha,\gamma\beta})(c_\gamma\psi_{\gamma,\gamma\beta})(b\psi_{\beta,\gamma\beta}) \\ &= a * c_\gamma * b = (a\rho) * b. \end{aligned}$$

Consequently $(\lambda,\rho) \in \Omega(S)$ which shows that $\lambda \in \tilde{\Lambda}(S)$. The first isomorphism in part i) follows by symmetry. Part ii) follows from part i) and 6.4. Finally part iii) follows easily from 6.4 and part i).

V.6.6 COROLLARY. If S is a (sturdy) semilattice of groups, then $\Omega(S)$ is also a (sturdy) semilattice of groups.

Proof. The statement above, excluding the words in parentheses, follows from 6.3 and 6.5. Let S be as in 6.4 with all $\psi_{\alpha,\beta}$ one-to-one and assume that $(a_\alpha)_{\alpha \in I}\, \varphi_{I,J} = (b_\alpha)_{\alpha \in I}\, \varphi_{I,J}$ in the notation of 6.3. Hence $a_\alpha = b_\alpha$ for all $\alpha \in J$. For every $\alpha \in I$ there exists $\beta \in J$ such that $I \cap J(\alpha) = J(\beta)$. Since then $a_\beta = b_\beta$, we must have $a_\alpha = b_\alpha$ for $\psi_{\alpha,\beta}$ is one-to-one. It follows that $(a_\alpha)_{\alpha \in I} = (b_\alpha)_{\alpha \in I}$ which shows that $\varphi_{I,J}$ is one-to-one.

V.6.7 COROLLARY. If S is a regular semigroup subdirect product of a semilattice and a group, then $\Omega(S)$ is also a regular semigroup subdirect product of a semilattice and a group.

Proof. This follows easily from IV.5.2 and 6.6.

V.6.8. Exercises

1.* Let S be a regular semigroup subdirect product of a semilattice Y and a group G. Show that $\Omega(S)$ is a subdirect product of $\mathcal{R}_Y$ and G. This sharpens 6.7.

2.* Let Y, L, G and R be a semilattice, a left zero semigroup, a group and a right zero semigroup, respectively. Prove
i) $\Lambda(Y \times L \times G \times R) \cong \mathfrak{I}(L)\; w\ell\, (G \times \Lambda(Y))$
where the wreath product on the right is formally defined in the same way as in 3.3,
ii) $\Omega(Y \times L \times G \times R) \cong \mathcal{R}_Y \times \mathfrak{I}(L) \times G \times \mathfrak{I}'(R)$.
(This generalizes 3.12.)

3. Let X be a nonempty set of cardinality η, let Q be the semigroup of all subsets of X under the set theoretic intersection, and for every $\tau \leq \eta$, let $Q_\tau = \{A \in Q \mid |A| < \tau\}$. Show that each Q_τ is a densely embedded ideal of Q for any $\tau > 2$.

4. Let Y be a finite semilattice and T be a nonempty subset of Y. Let $C_0 = T$ and for $n \geq 0$, suppose that C_n has been defined and let

$$C_{n+1} = \{\alpha \in Y \mid \alpha \leq \text{l.u.b. } \{\beta,\gamma\} \quad \text{for some} \quad \beta,\gamma \in C_n\}.$$

Show that $C = \bigcup_{n=0}^{\infty} C_n$ is the least retract ideal of Y containing T. Deduce that $\mathcal{R}_Y$ forms a lattice and find g.l.b. and l.u.b. in $\mathcal{R}_Y$. (For definitions of a greatest lower bound, g.l.b., and a least upper bound, l.u.b., see I.5.3.)

5.* Let (G,I) be an η-semigroup. Let m be a nonnegative integer and $\alpha \in G$ be such that $m + I(\alpha,\beta) \geq 1$ for all $\beta \in G$. Define a function $[m,\alpha]$ on (G,I) by

$$[m,\alpha](n,\beta) = (m + n + I(\alpha,\beta) - 1,\ \alpha\beta).$$

Prove that $[m,\alpha] \in \Lambda(S)$ and that, conversely, every $\lambda \in \Lambda(S)$ can be uniquely written in the form $[m,\alpha]$. Further show:
i) $[m,\alpha][n,\beta] = [m + n + I(\alpha,\beta) - 1,\ \alpha\beta]$,
ii) $\lambda_{(m,\alpha)} = [m + 1,\ \alpha]$,
iii) $[m,\alpha] \in \Gamma(S)$ if and only if $m > 0$,
iv) $[m,\alpha]$ is invertible if and only if $m = 0$ and $I(\alpha,\beta) = 1$ for all $\beta \in G$.

6. Show that in a semigroup S every element is its own unique inverse if and only if S is a semilattice of groups G_e in which $x^2 = e$ for all $x \in G_e$ Also show that such a semigroup is necessarily commutative.

7.* Prove that the following conditions on a semigroup S are equivalent.
   i) S is a chain of groups.
   ii) Every left and every right ideal of S is two-sided and completely prime.
   iii) S is completely regular and E_S is a chain.
   iv) For any $x,y \in S$, either $x \in yxSy$ or $y \in xySx$.

   Also show that if such a semigroup does not have an identity, then $\Omega(S) = \Pi(S) \cup \Sigma(S)$.

8. Show that the multiplicative semigroup of a ring R for which $x \in x^2R$ for every $x \in R$, is a semilattice of groups.

9. Let S be an $\mathfrak{N}$-semigroup and let D be the $\mathfrak{N}$-class of $\Omega(S)$ containing $\Pi(S)$. Show that D is an $\mathfrak{N}$-semigroup and an ideal of $\Omega(S)$, and that for every $\mathfrak{N}$-semigroup V which is an extension of S, there exists a unique isomorphism of V into D which extends the canonical homomorphism π of S into $\Omega(S)$.

V.6.9 REFERENCES: McMorris [1], Szász [1], Szász and Szendrei [1], Tamura [5], [14].

Appendix A

Semigroups of Order ≤ 4

We will not list all semigroups of order ≤ 4 but will proceed as follows. For any class $\mathcal{C}$ of semigroups, we may construct a possibly larger class by performing one or both of the following types of operations on the members of $\mathcal{C}$:

(α) adjunction of an identity or zero (even if they already have one), inflation,

(β) forming of direct products.

We will adhere to the following rules.

1. All isomorphs and antiisomorphs will be omitted.
2. For $n \leq 4$, semigroups of order n will be divided into 2 categories.

   i) Those that can be obtained from semigroups of order $< n$ by one of the above operations.

   ii) The remaining ones are listed by multiplication tables. We let

$$G_n = \text{the cyclic group of order } n,$$

$$R_n = \text{the right zero semigroup of order } n;$$

semilattices are represented by the associated lower semilattices. The rest of the semigroups are classified by the configuration of their greatest semilattice decompositions and the number of elements in each $\mathfrak{N}$-class, e.g., "$\{a,b,c\}$" means "one $\mathfrak{N}$-class, it has 3 elements"; further "$\{b,c\}$ d" means "the form of the

a

greatest semilattice homomorphic image is $\vee$ and one of the upper η-classes has 2 elements, the remaining ones 1 element each," etc. Multiplication tables contain only the products, but no variables, e.g.

$$\begin{matrix} a & a & a & a \\ a & a & a & a \\ a & a & a & b \\ a & a & b & c \end{matrix} \quad \text{stands for} \quad \begin{array}{c|cccc} & a & b & c & d \\ \hline a & a & a & a & a \\ b & a & a & a & a \\ c & a & a & a & b \\ d & a & a & b & c \end{array}$$

SEMIGROUPS OF ORDER 2

i) perform (α) on the semigroup of order 1.
ii) $\{a,b\}$: G_2, R_2

SEMIGROUPS OF ORDER 3

i) perform (α) on semigroups of order 2.
ii) $\{a,b,c\}$: G_3, R_3, $\begin{matrix} a & a & a \\ a & a & a \\ a & a & b \end{matrix}$

c over $\{a,b\}$: $\begin{matrix} a & a & a \\ a & a & a \\ a & b & c \end{matrix} \quad \begin{matrix} a & b & a \\ a & b & a \\ a & b & c \end{matrix}$

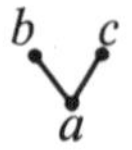

b, c over a

SEMIGROUPS OF ORDER 4

i) perform (α) on semigroups of order 3 and (β) on semigroups of order 2.

ii) $\{a,b,c,d\}$: G_4, $\begin{matrix} a & a & a & a \\ a & a & a & a \\ a & a & a & b \\ a & a & b & c \end{matrix} \quad \begin{matrix} a & a & a & a \\ a & a & a & a \\ a & a & a & a \\ a & a & b & b \end{matrix} \quad \begin{matrix} a & a & a & a \\ a & a & a & a \\ a & a & b & b \\ a & a & b & a \end{matrix}$

$$\begin{matrix} a & a & a & a \\ a & a & a & a \\ a & a & b & b \\ a & a & a & b \end{matrix} \quad \begin{matrix} a & a & a & a \\ a & a & a & a \\ a & a & b & a \\ a & a & a & b \end{matrix} \quad \begin{matrix} a & a & a & a \\ a & a & a & a \\ a & a & a & b \\ a & a & b & a \end{matrix} \quad \begin{matrix} a & a & a & a \\ a & a & a & a \\ a & a & a & a \\ a & a & b & a \end{matrix}$$

$$\begin{array}{cccc} a & b & a & a \\ b & a & b & b \\ a & b & a & a \\ a & b & a & c \end{array} \qquad \begin{array}{cccc} a & b & a & b \\ b & a & b & a \\ a & b & a & b \\ b & a & b & c \end{array} \qquad \begin{array}{cccc} a & b & c & a \\ a & b & c & a \\ a & b & c & b \\ a & b & c & a \end{array} \qquad \begin{array}{cccc} a & b & a & a \\ a & b & a & a \\ a & b & a & a \\ a & b & a & c \end{array}$$

$$\begin{array}{c} \{c,d\} \\ | \\ \{a,b\} \end{array} :$$

$$\begin{array}{cccc} a & b & a & a \\ a & b & a & a \\ a & b & c & d \\ a & b & c & d \end{array} \qquad \begin{array}{cccc} a & b & a & a \\ a & b & b & b \\ a & b & c & d \\ a & b & c & d \end{array} \qquad \begin{array}{cccc} a & a & a & a \\ a & a & a & a \\ a & b & c & d \\ a & b & c & d \end{array}$$

$$\begin{array}{cccc} a & a & a & a \\ a & a & b & b \\ a & a & c & d \\ a & a & c & d \end{array} \qquad \begin{array}{cccc} a & b & a & a \\ a & b & a & a \\ a & b & c & c \\ a & b & d & d \end{array} \qquad \begin{array}{cccc} a & b & a & a \\ a & b & b & b \\ a & b & c & c \\ a & b & d & d \end{array}$$

$$\begin{array}{cccc} a & b & a & a \\ a & b & b & b \\ a & b & c & d \\ a & b & d & c \end{array} \qquad \begin{array}{cccc} a & b & a & b \\ a & b & b & a \\ a & b & c & d \\ a & b & d & c \end{array} \qquad \begin{array}{cccc} a & b & b & b \\ a & b & b & b \\ a & b & c & d \\ a & b & d & c \end{array} \qquad \begin{array}{cccc} a & a & a & a \\ a & a & a & a \\ a & b & c & d \\ a & b & d & c \end{array}$$

$$\begin{array}{cccc} a & a & a & a \\ a & b & c & d \\ a & b & c & d \\ a & d & d & a \end{array} \qquad \begin{array}{cccc} a & b & a & a \\ b & a & b & b \\ a & b & c & d \\ a & b & c & d \end{array} \qquad \begin{array}{cccc} a & a & a & a \\ a & a & b & b \\ a & b & c & d \\ a & b & d & c \end{array} \qquad \begin{array}{cccc} a & b & a & a \\ b & a & b & b \\ a & b & c & d \\ a & b & d & c \end{array}$$

$$\begin{array}{c} d \\ | \\ \{a,b,c\} \end{array} :$$

$$\begin{array}{cccc} a & a & a & a \\ a & a & a & b \\ a & a & a & c \\ a & a & a & d \end{array} \qquad \begin{array}{cccc} a & a & a & a \\ a & a & a & b \\ a & a & a & a \\ a & a & c & d \end{array} \qquad \begin{array}{cccc} a & b & c & a \\ a & b & c & a \\ a & b & c & a \\ a & b & c & d \end{array}$$

$$\begin{array}{cccc} a & b & c & a \\ a & b & c & b \\ a & b & c & a \\ a & b & c & d \end{array} \qquad \begin{array}{cccc} a & b & a & a \\ a & b & a & a \\ a & b & a & c \\ a & b & a & d \end{array} \qquad \begin{array}{cccc} a & b & a & a \\ a & b & a & b \\ a & b & a & c \\ a & b & a & d \end{array}$$

$$\begin{array}{cccc} a & b & a & a \\ a & b & a & b \\ a & b & a & a \\ a & b & c & d \end{array} \qquad \begin{array}{cccc} a & b & a & a \\ a & b & a & a \\ a & b & a & a \\ a & b & c & d \end{array} \qquad \begin{array}{cccc} a & b & a & b \\ a & b & a & b \\ a & b & a & b \\ a & b & c & d \end{array} \qquad \begin{array}{cccc} a & b & b & b \\ a & b & b & b \\ a & b & b & c \\ a & b & c & d \end{array}$$

$$\begin{array}{cccc} a & a & a & a \\ a & a & a & b \\ a & a & a & c \\ a & b & a & d \end{array} \qquad \begin{array}{cccc} a & a & a & a \\ a & a & a & b \\ a & a & a & c \\ a & b & b & d \end{array} \qquad \begin{array}{cccc} a & b & a & a \\ b & a & b & b \\ a & b & a & a \\ a & b & c & d \end{array} \qquad \begin{array}{cccc} a & b & b & a \\ b & a & a & b \\ b & a & a & b \\ a & b & c & d \end{array}$$

$$\begin{matrix} a & a & a & a \\ a & a & a & a \\ a & a & a & a \\ a & b & b & d \end{matrix} \qquad \begin{matrix} a & a & a & a \\ a & a & a & a \\ a & a & b & a \\ a & a & a & d \end{matrix}$$

c d over $\{a,b\}$:

$$\begin{matrix} a & b & a & a \\ a & b & b & a \\ a & b & c & a \\ a & b & a & d \end{matrix} \qquad \begin{matrix} a & b & b & a \\ a & b & b & a \\ a & b & c & a \\ a & b & b & d \end{matrix} \qquad \begin{matrix} a & b & b & b \\ a & b & b & b \\ a & b & c & b \\ a & b & b & d \end{matrix}$$

$$\begin{matrix} a & a & a & a \\ a & a & b & a \\ a & a & c & a \\ a & a & a & d \end{matrix} \qquad \begin{matrix} a & a & a & a \\ a & a & b & a \\ a & a & c & a \\ a & b & a & d \end{matrix} \qquad \begin{matrix} a & a & a & a \\ a & a & b & a \\ a & a & c & a \\ a & b & b & d \end{matrix}$$

$$\begin{matrix} a & b & a & a \\ b & a & b & b \\ a & b & c & a \\ a & b & a & d \end{matrix} \qquad \begin{matrix} a & a & a & a \\ a & a & b & a \\ a & b & c & a \\ a & a & a & d \end{matrix}$$

$\{b,c\}$ d over a :

$$\begin{matrix} a & a & a & a \\ a & b & b & a \\ a & c & c & a \\ a & a & a & d \end{matrix} \qquad \begin{matrix} a & a & a & a \\ a & b & c & a \\ a & c & b & a \\ a & a & a & d \end{matrix}$$

d over c over $\{a,b\}$:

$$\begin{matrix} a & b & b & b \\ a & b & b & b \\ a & b & c & c \\ a & b & c & d \end{matrix} \qquad \begin{matrix} a & a & a & a \\ a & a & b & b \\ a & a & c & c \\ a & a & c & d \end{matrix} \qquad \begin{matrix} a & a & a & a \\ a & a & a & a \\ a & a & c & c \\ a & b & c & d \end{matrix}$$

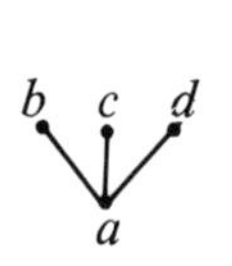

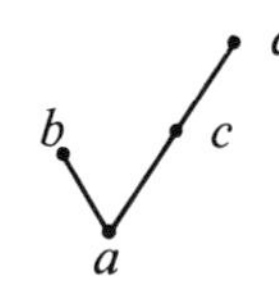

In the table below, the number of nonisomorphic and nonantiisomorphic semigroups of orders 2, 3 and 4 having the properties indicated in the first column is given.

order	2	3	4
i)	2	12	69
ii)	2	6	57
commutative	3	12	58
(completely) regular	3	9	42
inverse	2	5	16
1 idempotent	2	5	19
2 idempotents	2	7	37
3 idempotents		6	44
4 idempotents			26
total	4	18	126

Appendix B

List of Symbols

Italic

Boldface

German

Roman Script

Greek

Special Symbols

Bibliography

Adams, D.H.

[1] Semigroups with no non-zero nilpotent elements, *Math. Zeitschr.* 123 (1971), 168–176.

Allen, D.

[1] A generalization of the Rees theorem to a class of regular semigroups, *Semigroup Forum* 2(1971), 321–331.

Andrunakievič, V.A. and Rjabuhin, Ju. M.

[1] Rings without nilpotent elements and completely prime ideals, *Doklady Akad. Nauk SSSR* 180(1968), 9–11 (in Russian); Transl. *Soviet Math. Doklady* 9(1968), 565–568.

Arbib, M.I. (editor)

[1] *Algebraic theory of machines, languages, and semigroups*, Academic Press, New York, 1968.

Arendt, B.D. and Stuth, C.J.

[1] On the structure of commutative periodic semigroups, *Pacific J. Math.* 35(1970), 1–6.

[2] On partial homomorphisms of semigroups, *Pacific J. Math.* 35(1970), 7–9.

Birkhoff, G.

[1] *Lattice theory*, Amer. Math. Soc. Coll. Publ. No. 25, Providence, 1967 (3rd edition).

Brandt, H.

[1] Über eine Verallgemeinerung des Gruppenbegriffes, *Math. Ann.* 96 (1927), 360–366.

Brown, D.R. and LaTorre, J.G.

[1] A characterization of uniquely divisible commutative semigroups, *Pacific J. Math.* 18(1966), 57–60.

Bruck, R.H.

[1] *A survey of binary systems*, Springer, Berlin, 1958.

Burmistrovič, I.E.

[1] Commutative bands of cancellative semigroups, *Sibir. Mat. Ž.* 6(1965), 284–299 (in Russian).

Chevalley, C.

[1] *Fundamental concepts of algebra*, Academic Press, New York, 1956.

Chrislock, J.L.

[1] Semigroups whose regular representation is a right group, *Amer. Math. Monthly* 74(1967), 1097–1100.

[2] On medial semigroups, *J. Algebra* 12(1969), 1–9.

Clifford, A.H.

[1] Semigroups admitting relative inverses, *Annals of Math.* 42(1941), 1037–1049.

[2] Extensions of semigroups, *Trans. Amer. Math. Soc.* 68(1950), 165–173.

[3] Bands of semigroups, *Proc. Amer. Math. Soc.* 5(1954), 499–504.

[4] Radicals in semigroups, *Semigroup Forum* 1(1970), 103–127.

[5] The structure of orthodox unions of groups, *Semigroup Forum* 3(1972), 283–337.

Clifford, A.H. and Preston, G.B.

[1] *The algebraic theory of semigroups*, Math. Surveys No. 7, Amer. Math. Soc., Providence, Vol. I (1961), Vol. II (1967).

Croisot, R.

[1] Demi-groupes inversifs et demi-groupes réunions de demi-groupes simples, *Ann. Sci. Ecole Norm. Sup.* 70(1953), 361–379.

[2] Automorphismes intérieurs d'un semi-groupe, *Bull. Soc. Math. France* 82(1954), 161–194.

Dubreil, P.

[1] Contribution à la théorie des demi-groupes, *Mém. Acad. Sci. Inst. France* (2)63, no.3(1941), 52 pp.

[2] *Algèbre*, Vol. I, Gauthier-Villars, Paris, 1954 (2nd edition).

Dubreil-Jacotin, M.L., Lesieur, L., Croisot, R.

[1] *Leçons sur la théorie des treillis, des structures algébriques ordonnées et des treillis géometriques*, Gauthier-Villars, 1953.

Fantham, P.H.H.

[1] On the classification of a certain type of semigroup, *Proc. London Math. Soc.* (3)10(1960), 409–427.

Feller, E.H. and Gantos, R.L.

[1] Completely right injective semigroups that are unions of groups, *Glasgow Math. J.* 12(1971), 41–49.

Folley, K.W. (editor)

[1] *Semigroups*, Academic Press, New York, 1969.

Forsyth, G.E.

[1] SWAC computes 126 distinct semigroups of order 4, *Proc. Amer. Math. Soc.* 6(1955), 443–447.

Fuchs, L.

[1] *Partially ordered algebraic systems*, Pergamon, Oxford, 1963; German transl. *Teilweise geordnete algebraische Strukturen*, Vandehoeck-Ruprecht, Göttingen, 1966.

Gantos, R.L.

[1] Semilattices of bisimple inverse semigroups, *Quart. J. Math. Oxford* (2) 22(1971), 379–394.

Ginsburg, A.

[1] *Algebraic theory of automata*, Academic Press, New York, 1968.

Gluskin, L.M.

[1] Semigroups and rings of endomorphisms of linear spaces, *Izv. Akad. Nauk SSSR* 23(1959), 841–870 (in Russian); *Amer. Math. Soc. Transl.* 45(1965), 105–137.

[2] Ideals of semigroups of transformations, *Mat. Sbornik* 47(1959), 111–130 (in Russian).

[3] Ideals of semigroups, *Mat. Sbornik* 55(1961), 421–448; correction: ibid. 73(1967), 303 (in Russian).

[4] Semigroups of transformations, *Usp. Mat. Nauk* 17(1962), 233–240 (in Russian).

[5] On dense embeddings, *Mat. Sbornik* 61(1963), 175–206 (in Russian).

[6] Semigroups, *Itogi Nauki, Akad. Nauk SSSR* (1962), 33–58 (in Russian).

[7] ibid. (1964), 161–202 (in Russian).

[8] On separative semigroups, *Izv. Vysš. Učebn. Zaved. Mat.* 9(112) (1971), 30–39 (in Russian).

Gluskin, L.M., Schein, B.M. and Ševrin, L.N.

[1] Semigroups, *Itogi Nauki, Akad. Nauk SSSR* (1966), 9–56 (in Russian).

Goralčík, P.

[1] On translations of semigroups III. Transformations with increasing and transformations with irregular surjective part, *Mat. Časopis* 18(1968), 263–272 (in Russian).

Goralčík, P. and Hedrlín, Z.

[1] On translations of semigroups II. Surjective transformations, *Mat. Časopis* 18(1968), 263–272 (in Russian).

Green, J.A.

[1] On the structure of semigroups, *Annals of Math.* 54(1951), 163–172.

Grigor, R.S.

[1] Groupoids without nilpotent elements and zero associative groupoids, *Mat. Issl., Akad. Nauk MSSR* 5(1970), 40–55 (in Russian).

Grillet, M.P.

[1] On strongly strict extensions of semigroups, *J. Reine Angew. Math.* 246(1971), 169–171.

Grillet, P.A., and Grillet, M.P.

[1] On strict extensions of semigroups, *J. Reine Angew. Math.* 247(1971), 75–79.

Grillet, P.A. and Petrich, M.

[1] Ideal extensions of semigroups, *Pacific J. Math.* 26(1968), 493–508.

[2] Free products of semigroups amalgamating an ideal, *J. London Math. Soc.* (2) 2(1970), 389–392.

Hall, R.E.

[1] Commutative cancellative semigroups with two generators, *Czechoslovak Math. J.* 21(1971), 449–452.

Hall, T.E.

[1] On regular semigroups whose idempotents form a subsemigroup, *Bull. Austral. Math. Soc.* 1(1969), 195–208.

[2] On orthodox semigroups and uniform and antiuniform bands, *J. Algebra* 16(1970), 204–217.

Head, T.

[1] Commutative semigroups having greatest regular images, *Semigroup Forum* 2(1971), 130–137.

Hedrlín, Z. and Goralčík, P.

[1] On translations of semigroups I. Periodic and quasiperiodic transformations, *Mat. Časopis* 18(1968), 161–176 (in Russian).

Heuer, C.V.

[1] On the equivalence of cancellative extensions of commutative cancellative semigroups by groups, *Proc. Austral. Math. Soc.* 12(1971), 187–192.

Heuer, C.V. and Miller, D.W.

[1] An extension problem for cancellative semigroups, *Trans. Amer. Math. Soc.* 122(1966), 499–515.

Hewitt, E. and Zuckerman, H.S.

[1] The ℓ_1-algebra of a commutative semigroup, *Trans. Amer. Math. Soc.* 83(1956), 70–97.

Higgins, J.C.

[1] A faithful canonical representation of finitely generated $\mathfrak{N}$-semigroups, *Czechoslovak Math. J.* 19(1969), 375-379.

[2] Representing $\mathfrak{N}$-semigroups, *Bull. Austral. Math. Soc.* 1(1969), 115–125.

Hofmann, K.H. and Mostert, P.S.

[1] *Elements of compact semigroups*, C.E. Merrill Publishing Co., Columbus, Ohio, 1966.

Howie, J.M.

[1] Naturally ordered bands, *Proc. Glasgow Math. Assoc.* 8(1967), 55–58.

Howie, J.M. and Lallement, G.

[1] Certain fundamental congruences on a regular semigroup, *Proc. Glasgow Math. Assoc.* 7(1966), 145–156.

Ivan, J.

[1] On the decomposition of a simple semigroup into a direct product, *Mat.-Fyz. Časopis* 4(1954), 181–202 (in Slovak; Russian summary).

Johnson, B.E.

[1] An introduction to the theory of centralizers, *Proc. London Math. Soc.* (3) 14(1964), 299–320.

Kalmanovič, A.M.

[1] Densely embedded ideals of semigroups of multivalued partial endomorphisms of a graph, *Dopovidi Akad. Nauk URSR*, no.5(1967), 406–411 (in Ukrainian; Russian and English summaries).

Kapp, K.M. and Schneider, H.

[1] *Completely* 0-*simple semigroups* (an abstract treatment of the lattice of congruences), Benjamin, New York, Amsterdam, 1969.

Kaufman, A.M.

[1] Successively-annihilating sums of associative systems, *Uč. Zap. Leningrad. Gos. Ped. Inst.* 86(1949), 145–165 (in Russian).

Kimura, N.

[1] The structure of idempotent semigroups (I), *Pacific J. Math.* 8(1958), 257–275.

Kist, J.

[1] Minimal prime ideals in commutative semigroups, *Proc. London Math. Soc.* (3) 13(1963), 31–50.

Krohn, K., Rhodes, J. and Tilson, B.

[1] *Lectures on finite semigroups*, Vol. I, II, Univ. of Calif., Berkeley, 1967.

Lallement, G.

[1] Demi-groupes réguliers, *Ann. Mat. Pura Appl.* 77(1967), 47–130.

Lallement, G. and Petrich, M.

[1] Décompositions I-matricielles d'un demi-groupe, *J. Math. Pures Appl.* 45(1966), 67–117.

[2] Extensions of a Brandt semigroup by another, *Canad. J. Math.* 22(1970), 974–983.

Levin, R.G.

[1] On commutative, nonpotent archimedean semigroups, *Pacific J. Math.* 27(1968), 365–371.

Levin, R.G. and Tamura, T.

[1] Notes on commutative power joined semigroups, *Pacific J. Math.* 35(1970), 673–680.

Ljapin, E.S.

[1] Abstract characteristic of certain semigroups of transformations, *Uč. Zap. Leningrad. Gos. Ped. Inst.* 103(1955), 5–30 (in Russian).

[2] *Semigroups*, Fizmatgiz, Moscow, 1960 (in Russian); English transl. by Amer. Math. Soc., 1968 (2nd edition).

[3] On conditions for dense embedding of semigroups, *Uč. Zap. Leningrad. Gos. Ped. Inst.* 238(1962), 3–20 (in Russian).

Ljapin, E.S., Aizenštat, A. Ja. and Lesohin, M.M.

[1] *Exercises in group theory*, Nauka, Moscow, 1967 (in Russian); English transl. Plenum Press, New York, 1971.

McAlister, D.B.

[1] Characters on commutative semigroups, *Quart. J. Math. Oxford* (2) 19(1968), 141–157.

McAlister, D.B. and O'Carroll, L.

[1] Finitely generated commutative semigroups, *Glasgow Math. J.* 11(1970), 134–151.

McLean, D.

[1] Idempotent semigroups, *Amer. Math. Monthly* 6(1954), 110–113.

McMorris, F.R.

[1] The quotient semigroup of a semigroup that is a semilattice of groups, *Glasgow Math. J.* 12(1971), 18–23.

McNeil, P.E.

[1] Group extensions of null semigroups, *Duke Math. J.* 38(1971), 491–498.

Numakura, K.

[1] Note on the structure of commutative semigroups, *Proc. Japan Acad.* 30(1954), 262–265.

O'Carroll, L. and Schein, B.M.

[1] On exclusive semigroups, *Semigroup Forum* 3(1972), 338–348.

Paalman-de Miranda, A.B.

[1] *Topological semigroups*, Mat. Centre tracts no. 11 Amsterdam, 1964.

Papy, G.

[1] *Groupoïdes*, Presses Univ., Paris, 1965.

Petrich, M.

[1] Idéaux demi-premiers et premiers du produit cartésien d'un nombre fini de demi-groupes, *C.R. Acad. Sci., Paris* 256(1963), 3950–3953.

[2] The maximal semilattice decomposition of a semigroup, *Math. Zeitschr.* 85(1964), 68–82.

[3] On the structure of a class of commutative semigroups, *Czechoslovak Math. J.* 14(1964), 147–153.

[4] Semigroups certain of whose subsemigroups have identities, *Czechoslovak Math. J.* 16(1966), 186–198.

[5] On extensions of semigroups determined by partial homomorphisms, *Nederl. Akad. Wetensch. Indag. Math.* 28(1966), 49–51.

[6] Homomorphisms of a semigroup onto normal bands, *Acta Sci. Math. Szeged* 27(1966), 185–196.

[7] Sur certaines classes de demi-groupes, III, *Acad. Roy. Belg. Cl. Sci.* 53(1967), 60–73.

[8] Congruences on extensions of semigroups, *Duke Math. J.* 34(1967), 215–224.

[9] *Topics in semigroups*, Pennsylvania State Univ. 1967.

[10] Inflation of a completely 0-simple semigroup, *Bull. Soc. Math. Belg.* 19(1967), 42–54.

[11] The translational hull of a completely 0-simple semigroup, *Glasgow Math. J.* 9(1968), 1–11.

[12] Translational hull and semigroups of binary relations, *Glasgow Math. J.* 9(1968), 12–21.

[13] The semigroup of endomorphisms of a linear manifold, *Duke Math. J.* 36(1969), 145–152.

[14] Representations of semigroups and the translational hull of a regular Rees matrix semigroup, *Trans. Amer. Math. Soc.* 143(1969), 303–318.

[15] *Semigroups and rings of linear transformations*, Pennsylvania State Univ., 1969.

[16] The translational hull in semigroups and rings, *Semigroup Forum* 1(1970), 283–360.

[17] Certain dense embeddings of regular semigroups, *Trans. Amer. Math. Soc.* 155(1971), 333–343.

Petrich, M. and Grillet, P.A.

[1] Extensions of an arbitrary semigroup, *J. Reine Angew. Math.* 244(1970), 97–107.

Plemmons, R.J.

[1] *Cayley tables for all semigroups of order* ≤ 6, Auburn Univ., Alabama, 1965.

Pondělíček, B.

[1] A certain equivalence on a semigroup, *Czechoslovak Math. J.* 21(1971), 109–117.

Ponizovskiĭ, I.S.

[1] A remark on inverse semigroups, *Usp. Mat. Nauk* 20(1965), 147–148 (in Russian).

Preston, G.B.

[1] The structure of normal inverse semigroups, *Proc. Glasgow Math. Assoc.* 3(1956), 1–9.

Putcha, M.S. and Weissglass, J.

[1] Semigroups satisfying variable identities, *Semigroup Forum* 3(1971), 64–67.

Rédei, L.

[1] *Theorie der endlich erzeugbaren kommutativen Halbgruppen*, Physica

Verlag, Würzburg, 1963; English transl., *The theory of finitely generated commutative semigroups*, Pergamon Press, Oxford, New York, 1965.

Rees, D.

[1] On semi-groups, *Proc. Cambridge Phil. Soc.* 36(1940), 387–400.

Rosenfeld, A.

[1] *An introduction to algebraic structures*, Holden-Day, San Francisco, 1968.

Saitô, T. (editor)

[1] *Memoirs of a seminar on the algebraic theory of semigroups*, Res. Inst. Math. Sci., Kyoto, 1967.

Schein, B.M.

[1] On the theory of restrictive semigroups, *Izv. Vysš. Učebn. Zaved. Mat.* 2(33) (1963), 152–154 (in Russian).

[2] On translations in semigroups and groups, *Volž. Mat. Sbornik* 2(1964), 163–169 (in Russian).

[3] On a class of commutative semigroups, *Publ. Math. Debrecen* 12(1965), 87–88 (in Russian).

[4] Homomorphisms and subdirect decompositions of semigroups, *Pacific J. Math.* 17(1966), 529–547.

[5] *Lectures in transformation semigroups*, Izdat. Saratov. Univ., 1970 (in Russian).

[6] A note on radicals in regular semigroups, *Semigroup Forum* 3(1971), 84–85.

Schwarz, Š.

[1] Contribution to the theory of periodic semigroups, *Czechoslovak Math. J.* 3(1953), 139–153 (in Russian).

[2] Semigroups in which every proper subideal is a group, *Acta Sci. Math. Szeged* 21(1960), 125–134.

Sedlock, J.T.

[1] Green's relations on a periodic semigroup, *Czechoslovak Math. J.* 19(1969), 318–323.

Selfridge, J.L.

[1] *On finite semigroups* (Appendix: semigroups of order 5), Doctoral Dissertation, Univ. of Calif., Los Angeles, 1958.

Skornjakov, L.A.

[1] Left chained semigroups, *Sibir. Mat. Ž.* 11(1970), 168–182 (in Russian).

Steinfeld, O.

[1] On semigroups which are unions of completely 0-simple semigroups, *Czechoslovak Math. J.* 16(1966), 63–69.

Suškevič, A.K. (Suschkewitsch)

[1] Über die endlichen Gruppen ohne das Gesetz der eindeutigen Umkehrbarkeit, *Math. Ann.* 99(1928), 30–50.

[2] *The theory of generalized groups*, Gos. Naučno-Tehn. Izd. Ukraini, Harkov, Kiev, 1937 (in Russian).

Szász, G.

[1] Die Translationen der Halbverbände, *Acta Sci. Math. Szeged* 17(1956), 165–169.

[2] Eine Charakteristik der Primidealhalbgruppen, *Publ. Math. Debrecen* 17(1970), 209–213.

Szász, G. and Szendrei, J.

[1] Über die Translationen der Halbverbände, *Acta Sci. Math. Szeged* 18(1957), 44–47.

Ševrin, L.N.

[1] On densely embedded ideals of semigroups, *Doklady Akad. Nauk SSSR* 131(1960), 765–768; correction: ibid. 164(1965), 1214 (in Russian); Transl. *Soviet Math. Doklady* 1(1960), 348–351.

[2] Completely simple semigroups without zero and idealizers of semigroups, *Izv. Vysš. Učebn. Zaved. Mat.* 6(55)(1966), 157–160.

[3] Densely embedded ideals of semigroups, *Mat. Sbornik* 79(1969), 425–432 (in Russian); Transl. *Math. USSR Sbornik* 8(1969, 401–408.

Širjaev, V.M.

[1] On the semigroup of pairs of inner translations of a semigroup, *Vesci Akad. Nauk BSSR*, Ser. fiz-mat. nauk, no. 4(1969, 99–107 (in Russian).

Šulka, R.

[1] The maximal semilattice decomposition of a semigroup, radicals and nilpotency, *Mat. Časopis* 20(1970), 172–180.

Šutov, E.G.

[1] On translations of semigroups, *Usp. Mat. Nauk* 19(1964), 215–216 (in Russian).

Tamura, T.

[1] Note on finite semigroups and determination of semigroups of order 4, *J. Gakugei, Tokushima Univ.* 5(1954), 17–28.

[2] One-sided bases and translations of a semigroup, *Math. Japonica* 3(1955), 137–141.

[3] On translations of a semigroup, *Kōdai Math. Sem. Rep.* 7(1955), 67–70.

[4] The theory of construction of finite semigroups I, *Osaka Math. J.* 8(1956), 243–261.

[5] The theory of construction of finite semigroups II, *Osaka Math. J.* 9(1957), 1–42.

[6] Commutative nonpotent archimedean semigroup with cancellation law. I., *J. Gakugei, Tokushima Univ.* 8(1957), 5–11.

[7] Notes on translations of a semigroup, *Kōdai Math. Sem. Rep.* 10(1958), 9–26.

[8] Another proof of a theorem concerning the greatest semilattice decomposition of a semigroup, *Proc. Japan Acad.* 40(1964), 777–780.

[9] Attainability of systems of identities on semigroups, *J. Algebra* 3(1966), 261–276.

[10] Decomposability of extension and its application to finite semigroups, *Proc. Japan Acad.* 43(1967), 93–97.

[11] Maximal or greatest homomorphic images of given type, *Canad. J. Math.* 20(1968), 264–271.

[12] Construction of trees and commutative archimedean semigroups, *Math. Nachr.* 36(1968), 255–287.

[13] The study of closets and free contents related to semilattice decomposition of semigroups, in Folley [1], 221–259.

[14] Semigroups satisfying identity $xy = f(x,y)$, *Pacific, J. Math.* 31(1969), 513–521.

[15] Finite union of commutative power joined semigroups, *Semigroup Forum* 1(1970), 75–83.

[16] On commutative exclusive semigroups, *Semigroup Forum* 2(1971), 181–187.

Tamura, T. and Graham, N.

[1] Certain embedding problems for semigroups, *Proc. Japan Acad.* 40(1964), 8–13.

Tamura, T. and Kimura, N.

[1] On decompositions of a commutative semigroup, *Kōdai Math. Sem. Rep.* 6(1954), 109–112.

[2] Existence of greatest decomposition of a semigroup, *Kōdai Math. Sem. Rep.* 7(1955), 83–84.

Tamura T. et al.

[1] All semigroups of order at most 5, *J. Gakugei, Tokushima Univ.* 6(1955), 19–39.

Thierrin, G.

[1] Quelques propriétés des sous-groupoïdes consistants d'un demi-groupe abélien, *C.R. Acad. Sci. Paris* 236(1953), 1837–1839.

[2] Sur quelques propriétés de certaines classes de demi-groupes, *C.R. Acad. Sci., Paris* 239(1954), 1335–1337.

[3] Sur la théorie des demi-groupes, *Comment, Math. Helv.* 30(1956), 211–223.

[4] Sur quelques décompositions des groupoïdes, *C.R. Acad. Sci., Paris* 242(1956), 596–598.

[5] Sur les automorphismes intérieurs d'un demi-groupe réductif, *Comment. Math. Helv.* 31(1956), 145–151.

[6] Sur la structure des demi-groupes, *Alger Math.* 3(1956), 161–171.

Tully, J.R.

[1] Semigroups in which each ideal is a retract, *J. Austral. Math. Soc.* 9(1969), 239–245.

Vagner, V.V. (editor)

[1] *The theory of semigroups and its applications*, Izdat. Saratov. Univ., 1965 (in Russian).

Venkatesan, P.S.

[1] On decomposition of semigroups with zero, *Math. Zeitschr.* 92(1966), 164–174.

Verbeek, L.A.M.

[1] Union extensions of semigroups, *Trans. Amer. Math. Soc.* 150(1970), 409–423.

Warne, R.J.

[1] Extensions of completely 0-simple semigroups by completely 0-simple semigroups, *Proc. Amer. Math. Soc.* 17(1966), 524–526.

[2] Extensions of Brandt semigroups and applications, *Illinois J. Math.* 10(1966), 652–660.

Williamson, J.H.

[1] Harmonic analysis on semigroups, *J. London Math. Soc.* 42(1967), 1–41.

Yamada, M.

[1] On the greatest semilattice decomposition of a commutative semigroup, *Kōdai Math. Sem. Rep.* 7(1955), 59–62.

[2] Compositions of semigroups, *Kōdai Math. Sem. Rep.* 8(1956), 107–111.

[3] A remark on periodic semigroups, *Sci. Rep., Shimane Univ.* 9(1959), 1–5.

[4] A note on subdirect decompositions of idempotent semigroups, *Proc. Japan Acad.* 36(1960), 411–414.

[5] Strictly inversive semigroups, *Bull. Shimane Univ.* 13(1964), 128–138.

[6] Construction of finite commutative semigroups, *Bull. Shimane Univ.* 15(1965), 1–11.

[7] Regular semigroups whose idempotents satisfy permutation identities, *Pacific J. Math.* 21(1967), 371–392.

[8] Commutative ideal extensions of null semigroups, *Mem. Fac. Lit. Sci., Shimane Univ.* 1(1968), 8–22.

[9] On a regular semigroup in which the idempotents form a band, *Pacific J. Math.* 33(1970), 261–272.

Yamada, M. and Tamura, T.

[1] Note on finite commutative nil-semigroups, *Portugaliae Math.* 28(1969), 189–203.

Yoshida, R.

[1] Ideal extensions of semigroups and compound semigroups, *Mem. Res. Inst. Sci. Eng., Ritumeikan Univ.* 13(1965), 1–8.

Yoshida, R. and Yamada, M.

[1] On commutativity of a semigroup which is a semilattice of commutative semigroups, *J. Algebra* 11(1969), 278–297.

Yoshida, R. et al.

[1] Remarks on finite commutative z-semigroups, *Mem. Res. Inst. Sci. Eng., Ritumeikan Univ.* 16(1967), 1–11.

[2] On chains of semigroups, *Mem. Res. Inst. Sci. Eng., Ritumeikan Univ.* 17(1968), 1-10.

Zupnik, D.

[1] Cayley functions, *Semigroup Forum* 4(1972), 349–358.

Index